visual education

581 West Leffel Lane
P.O. Box 1666
Springfield, Ohio 45501
1-937-328-2854 DIRECT
1-800-243-7070 U.S.A.
1-937-324-5697 FAX
sales@vis-ed.com - E-mail
www.vis-ed.com - Web Page

VIS-ED®

a leading publisher in academics

ENGLISH

Think English:	Level 1, Cards & Cassette (ESL & Remedial)
	Level 2, Cards & Cassette (ESL & Remedial)
Vis-Ed Cards:	English Vocabulary Set 1
	English Vocabulary Set 2
	Spelling
Compact Facts:	Composition
	Grammar
	Punctuation
	Public Speaking (Speech)

ARABIC

Vis-Ed Cards:	Arabic Vocabulary

CHINESE

Vis-Ed Cards:	Chinese Vocabulary

FRENCH

Think French:	Level 1, Cards & Cassette
	Level 2, Cards & Cassette
Vis-Ed Cards:	French Vocabulary
Compact Facts:	Conversation
	Grammar
	Verbs

GERMAN

Think German:	Level 1, Cards & Cassette
	Level 2, Cards & Cassette
Vis-Ed Cards:	German Vocabulary
Compact Facts:	Conversation
	Grammar
	Verbs

GREEK & HEBREW

Vis-Ed Cards:	Biblical Greek Vocabulary
	Classical Greek Vocabulary
	Biblical Hebrew Vocabulary
	Modern Hebrew Vocabulary

ITALIAN

Vis-Ed Cards:	Italian Vocabulary
Compact Facts:	Conversation

JAPANESE

Vis-Ed Cards:	Japanese Vocabulary

KOREAN

Vis-Ed Cards:	Korean Vocabulary

LATIN

Vis-Ed Cards:	Latin Vocabulary

PORTUGUESE

Vis-Ed Cards:	Portuguese Vocabulary

RUSSIAN

Think Russian:	Level 1, Cards & Cassette
Vis-Ed Cards:	Russian Vocabulary
Compact Facts:	Conversation
	Grammar
	Verbs

RELIGION

Vis-Ed Cards:	Bible Verses, King James Version
	Bible Verses, New International Version

1.1 PLANE GEOMETRY: DEFINITIONS

What is geometry?

1.1

That branch of mathematics dealing with the properties, measurements and relations of lines, angles, surfaces and solids. The term is derived from two Greek words meaning ''earth measure''.

1.2 PLANE GEOMETRY: DEFINITIONS

What is the difference between plane and solid geometry?

1.2

Plane geometry deals with the relations of lines, angles and shapes in two dimensions; solid geometry extends these relations to three dimensions.

1.3 PLANE GEOMETRY: DEFINITIONS

What is Euclidean Geometry?

1.3

A Geometry developed by the Greeks, and summarized in Euclid's Books written about 300 B. C. It is based on a set of axioms, such as the axiom that the shortest distance between two points is a straight line, and uses the synthetic method of proof: that is, reasoning from assumed axioms or proven propositions to new conclusions.

1.4 PLANE GEOMETRY: DEFINITIONS

What is Analytic Geometry?

1.4

A geometry developed much later than the Euclidean, which treats geometrical problems from an algebraic point of view, through the media of Cartesian coordinates.

1.5 PLANE GEOMETRY: DEFINITIONS

What is an axiom?

1.5

A statement accepted as a truth without proof. Any synthetic type geometry has to begin with certain axioms which are accepted as self evident truths. For example, a fundamental axiom of Euclidean Geometry is "The whole is greater than any one of its parts".

1.6 PLANE GEOMETRY: DEFINITIONS

What is a postulate?

1.6

A postulate is also something assumed as true; or a statement that something is possible to be done, as a construction. For example, the statement that a circle can be constructed with any center and any radius is a postulate.

1.7 PLANE GEOMETRY: DEFINITIONS

What is a point?

1.7

A point, geometrically speaking, is something having position but without extent; or that which generates a line; or the intersection of two lines.

What is a line?

1.8

A line, geometrically speaking, is that which has length but not breadth; or the path of a moving point; or the intersection of two planes.

1.9 PLANE GEOMETRY: SYMBOLS

What do the symbols $\angle$ and $\triangle$ represent?

1.9

$\angle$ represents angle.
▲ represents angles.

PLANE GEOMETRY: DEFINITIONS

What is an angle?

1.10

An angle is the space within two lines which meet at a point called the vertex. This is a plane angle. When two planes intersect, the angles formed are called dihedral angles.

1.11 PLANE GEOMETRY: DEFINITIONS

How is an angle measured?

1.11

An angle is measured usually by one of two systems: 1) sexagesimal system - one revolution represents 360 degrees 2) radian system - one revolution represents 2π radians.

What is a straight angle and what is its measure?

1.12

A straight angle is an angle whose sides lie in a straight line; its measure is 180° or π radius.

1.13 PLANE GEOMETRY: DEFINITIONS

What is a right angle and what is its measure?

1.13

If one straight line meets another so as to make the two angles formed equal, each of the angles is a right angle. Its measure is 90° or $\pi/2$ radians.

1.14 PLANE GEOMETRY: SYMBOLS

What does the symbol $\perp$ represent?

1.14

$\perp$ represents perpendicular; two lines meeting at right angles are perpendicular to each other.

1.15 PLANE GEOMETRY: DEFINITIONS

What is an acute angle?

1.15

An acute angle is any angle smaller than a right angle, or 90°.

What is an obtuse angle?

1.16

An obtuse angle is any angle larger than a right angle but smaller than a straight angle; its measure is greater than 90° and less than 180°.

1.17 PLANE GEOMETRY: DEFINITIONS

What are opposite or vertical angles?

1.17

The angles opposite to each other when two straight lines cross are called opposite or vertical angles.

x and y are opposite angles.
p and q are opposite angles.

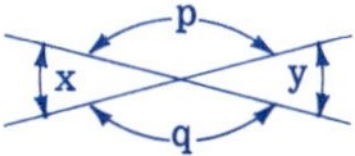

What are adjacent angles?

1.18

Adjacent angles are two angles which have the same vertex and have a common side between them.

What are supplementary angles?

1.19

Supplementary angles are two angles whose combined measure is a straight angle or 180°.

What are complementary angles?

1.20

Complementary angles are two angles whose combined measure is a right angle or 90°.

What is a circle?

1.21

A circle is a closed plane curve, all points of which are the same distance from a fixed point in the plane called the center. The length of this closed curve is called the circumference.

What is an arc of a circle?

1.22

Any part of the circumference of the circle is called an arc of the circle.

1.23 PLANE GEOMETRY: SYMBOLS

What does the symbol $\frown$ represent?

1.23

$\frown$ represents "arc."

What is a chord of a circle?

1.24

A chord of a circle is a straight line joining two points of the circle.

What is a tangent to a circle?

1.25

A tangent is a line which touches the circle at one, and only one, point.

What is a secant of a circle?

1.26

A secant is a line which intersects the circle in two points.

What is a diameter of a circle?

1.27

A diameter is any chord of the circle through the center of the circle.

What is an inscribed angle of a circle?

1.28

An inscribed angle is an angle formed by two chords meeting on the circle.

What are concentric circles?

1.29

Concentric circles are circles having the same center.

What is meant by bisect?

1.30

To bisect is to cut into two equal parts or halves.

What are congruent figures?

1.31

Figures which can be made to coincide are congruent. This means all the parts (sides or angles) of one equal respectively the corresponding parts of the other.

What does the symbol $\cong$ represent?

1.32

$\cong$ represents "congruent".

What are parallel lines?

1.33

Parallel lines, in Euclidean geometry, are straight lines in the same plane, that cannot meet, however far extended in either direction.

What is a transversal?

1.34

A transversal is a line that cuts two or more other lines.

PLANE GEOMETRY: SYMBOLS

What does the symbol $//$ represent?

1.35

// represents "parallel"

1.36 PLANE GEOMETRY: SYMBOLS

What does the symbol $\backsim$ represent?

1.36

$\backsim$ represents "similar."

What is a polygon?

1.37

A polygon is a closed figure formed by, or bounded by, straight lines.

What is meant by the perimeter of a polygon?

1.38

The perimeter of a polygon is a measure of the total length of all the sides of the polygon.

1.39 PLANE GEOMETRY: DEFINITIONS

What is a triangle?

1.39

A triangle is a polygon having three sides.

1.40 PLANE GEOMETRY: SYMBOLS

What does the symbol $\triangle$ and $\blacktriangle$ represent?

1.40

△ represents "triangle."

▲ represents "triangles."

What is an isosceles triangle?

1.41

An isosceles triangle is a triangle having at least two of its sides equal.

1.42 PLANE GEOMETRY: DEFINITIONS

What is an equilateral triangle?

1.42

An equilateral triangle is a triangle having all three sides equal.

What is a right triangle?

1.43

A right triangle is a triangle one of whose angles is a right angle. The side opposite the right angle is called the hypotenuse.

What is an altitude of a triangle?

1.44

An altitude of a triangle is a line drawn from any vertex $\perp$ to the opposite side (extended if necessary).

What is a median of a triangle?

1.45

A median of a triangle is a line drawn from any vertex to the midpoint of the opposite side.

What is an angle bisector of a triangle?

1.46

An angle bisector of a triangle is a line bisecting an angle of a triangle and extending to the opposite side.

What is a quadrilateral?

1.47

A quadrilateral is a four-sided polygon.

What is a trapezoid?

1.48

A trapezoid is a quadrilateral having two, and only two, sides parallel.

What is a parallelogram?

1.49

A parallelogram is a quadrilateral whose pairs of opposite sides are parallel.

What is a rhombus?

1.50

A rhombus is a parallelogram whose sides are all equal, but whose angles are not right angles.

What is a rectangle?

1.51

A rectangle is a parallelogram all of whose angles are right angles.

What is a square?

1.52

A square is a parallelogram whose sides are all equal and whose angles are all right angles.

What is a diagonal of a quadrilateral?

1,53

A diagonal of a quadrilateral is a line joining opposite vertices.

What is a regular polygon?

1.54

A regular polygon is a polygon which is both equilateral and equiangular.

What is meant by ratio and proportion?

1.55

A ratio is obtained by dividing one quantity by another; a proportion is the equality of two ratios.

In the proportion $\frac{a}{b}=\frac{c}{d}$, which quantities are the means?

1.56

b and c are the means: if $b = c$, then b is called the mean proportional to a and d.

1.57 PLANE GEOMETRY: SYMBOLS

What do the symbols $>$ and $<$ represent?

1.57

>represents "greater than."
<represents "less than."

2.1 TRIANGLES: CONGRUENT

What determines the congruency of two triangles?

2.1

Two triangles are congruent if:

1) Two angles and the included side of one equal respectively two angles and the included side of the other.
2) Two sides and the included angle of one equal respectively two sides and the included angle of the other.
3) Three sides of one are equal respectively to three sides of the other.

2.2 TRIANGLE: SIDES

What can be said about the sum and difference of two sides of a triangle with respect to the third side?

2.2

The sum of any two sides of a triangle is greater than the third side, and their difference is less than the third side.

TRIANGLE

If two sides of a triangle are equal, what can be said about the angles opposite these sides?

2.3

They are also equal.

2.4 TRIANGLE

If two angles of a triangle are equal, what can be said about the sides opposite these angles?

2.4

They are also equal.

2.5 TRIANGLE: ⊥ BISECTORS

The ⊥ bisectors of the sides of a triangle meet in a point called ________?

2.5

The center of the circumscribed circle.

2.6 TRIANGLE: ANGLE BISECTORS

The angle bisectors of a triangle meet in a point called ________?

2.6

Center of inscribed circle.

2.7 TRIANGLE: MEDIAN

The medians of a triangle meet in a point called ________?

2.7

The centroid. This point is two-thirds of the distance from any vertex to the midpoint of the opposite side.

2.8 PYTHAGOREAN THEOREM

What is the theorem of Pythagoras?

2.8

In any right triangle the square of the hypotenuse is equal to the sum of the squares of the other two sides.

2.9 TRIANGLE: CONGRUENCY

What determines the congruency of two right triangles?

2.9

Two right triangles are congruent if 1) a side and the hypotenuse of one are equal respectively to a side and the hypotenuse of the other 2) Same for the hypotenuse and an actue angle 3) Same for the two legs 4) Same for a side and the adjacent angle.

2.10 TRIANGLES: SIMILAR

What determines the similarity of two triangles?

2.10

Two triangles are similar if 1) two angles of one are equal to two angles of another 2) an angle of one is equal to an angle of the other and the sides which include these angles are proportional 3) their sides are proportional 4) their sides are respectively parallel 5) their sides are respectively perpendicular.

2.11 TRIANGLE: ISOSCELES

What determines an isosceles triangle?

2.11

A triangle is isosceles if
1) two sides are equal or
2) two angles are equal

2.12 LINES: TRANSVERSAL

If two parallel lines are cut by a transversal, which pairs of angles are equal?

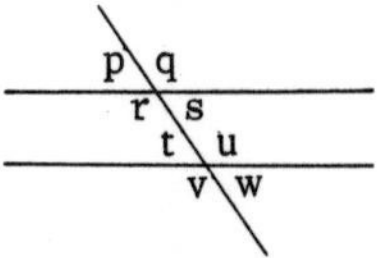

2.12

1) $p = t$, $q = u$, etc. (corresponding angles)
2) $p = s$, $q = r$, etc. (opposite or vertical angles)
3) $r = u$, $s = t$, (alternate interior angles)
4) $q = v$, $p = w$ (alternate exterior angles)

2.13 LINES: PARALLEL

What is the "parallel postulate?"

2.13

The parallel postulate states that through a given external point one and only one straight line can be drawn parallel to a given straight line.

LINES: PARALLEL

What determines the parallelism of two lines?

2.14

Two lines are parallel if;
1) Alternate interior angles are equal
2) Corresponding angles are equal
3) They are parallel to the same line
4) They are perpendicular to the same line
5) They are opposite sides of a parallelogram.

2.15 LINES: PERPENDICULAR

What determines the perpendicularity of two lines?

2.15

Two lines are perpendicular if:

1) Two points on one are equally distant from two points on the other.
2) They intersect so as to form two supplementary adjacent angles that are equal.
3) They are the sides of a triangle in which the sum of two angles equals the third angle.

2.16 ANGLES; SIDES OF

If two angles have their sides respectively

a) parallel, the angles are ________.

b) perpendicular, the angles are ________.

2.16

a) equal or supplementary.
b) equal or supplementary.

2.17 QUADRILATERAL

What determines whether a quadrilateral is a parallelogram?

2.17

1) opposite sides are parallel, or
2) opposite sides are equal, or
3) two sides are both equal and parallel, or
4) diagonals bisect each other

2.18 TRIANGLE

If a line bisects one side of a triangle and is parallel to a second side, it ________ the third side.

2.18

Bisects.

TRIANGLE: PROPERTIES

The line joining the midpoints of any two sides of a triangle is ________ to the third side and equal to ______ of the third side.

2.19

parallel, one-half.

2.20 TRIANGLE - INTERIOR ANGLES

What is the sum of the three interior angles of a triangle?

2.20

The sum of the three interior angles of a triangle is 180°.

2.21 TRIANGLE - ANGLE MEASURE

In a triangle ABC, find the angle C if $C = A + B$.

2.21

$$A + B + C = 180^\circ$$
$$A + B - C = 0^\circ$$

Subtracting,

$$2C = 180^\circ$$
$$C = 90^\circ$$

2.22 SQUARE

a) The area of a square of side S is ________,
b) the perimeter is ______,
c) the length of the diagonal is ______.

2.22

a) s^2
b) $4s$
c) $s\sqrt{2}$

2.23 RECTANGLE

The area of a rectangle of sides a and b is (1)______, the perimeter is (2)______, the length of a diagonal is (3)______?

2.23

(1) ab

(2) $2(a+b)$

(3) $\sqrt{a^2 + b^2}$

TRIANGLE: RIGHT

The area of a right triangle with legs a and b is _______?

2.24

$$\frac{ab}{2}$$

2.25 TRIANGLE: ISOSCELES

The area of a right isosceles triangle with legs each equal to a is ________, the perimeter is ________?

2.25

$$\frac{a^2}{2}$$

$$(2+\sqrt{2})\,a$$

2.26 TRIANGLE: EQUILATERAL

The area of an equilateral triangle of side a is ________, the perimeter is ________, an altitude is of length ______?

2.26

$$\frac{\sqrt{3}\,a^2}{4}$$

$$3a$$

$$\frac{\sqrt{3}}{2}a$$

TRIANGLE: RIGHT

The area of a 30°, 60° right triangle with hypotenuse a is ________, the perimeter is ________?

2.27

$$\frac{\sqrt{3}\,a^2}{8}$$

$$(3+\sqrt{3})\frac{a}{2}$$

2.28 TRIANGLE AREA

The area of an oblique triangle with sides a, b and c is
________?

2.28

If $s = \frac{a+b+c}{2}$, the area is $\sqrt{s(s-a)(s-b)(s-c)}$.

2.29 CIRCLE: INSCRIBED

What is the radius of a circle inscribed in a triangle?

2.29

$$\text{The radius} = \frac{2\ (\text{area of triangle})}{\text{perimeter of triangle}}.$$

TRIANGLE: ISOSCELES

The base of an isosceles triangle is 12 in., and each of the equal sides is 10 in. Find the area of the inscribed circle.

2.30 $$r = \frac{2(\text{area of the triangle})}{(\text{perimeter of the triangle})}$$

Area of $\Delta = \sqrt{s(s-a)(s-b)(s-c)}$ $\quad s = \frac{a+b+c}{2} = \frac{32}{2} = 16$

Area of $\Delta = \sqrt{16 \cdot 4 \cdot 6 \cdot 6} = 4 \cdot 2 \cdot 6 = 48$ sq. in.

$$\therefore r = \frac{2 \cdot 48}{12+2(10)} = \frac{2 \cdot 48}{32} = 3 \text{ in.}$$

Area of circle $= \pi r^2 = 9\pi$ sq. in. $= 28.27$ sq. in.

A square is surmounted by an equilateral triangle of perimeter 36. Find the area of the complete figure.

2.31
side s of $\Delta = \frac{36}{3} = 12$. Therefore, side of square $= s = 12$.
Area of square $= s^2 = 144$. Area of triangle $= \frac{\sqrt{3}\,s^2}{8} = 18\sqrt{3}$.

Total area $= 36 + 18\sqrt{3}$.

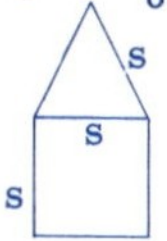

2.32 PARALLELOGRAM

The area of a parallelogram with one side b and altitude on that side h is ________.

2.32

bh

2.33

TRAPEZOID

The area of a trapezoid with parallel sides a and b and distance between parallel sides h is ________?

2.33

$$\frac{h\ (a + b)}{2}$$

2.34

RHOMBUS

The area of a rhombus with diagonals d_1 and d_2 is _____?

2.34

$$\frac{1}{2} d_1 d_2$$

RHOMBUS

The sum of the diagonals of a rhombus is 28 in. If the difference between the diagonals is 4 in., find the area of the rhombus.

2.35 Area $= \frac{d_1 \cdot d_2}{2}$

$$\begin{aligned} d_1 + d_2 &= 28 \\ d_1 - d_2 &= 4 \\ \hline 2d_1 &= 32 \text{ by addition.} \end{aligned}$$

$\therefore d_1 = 16$ in. ; $d_2 = 12$ in.

$\therefore$ Area $= \frac{16 \cdot 12}{2} = 96$ sq. in.

2.36 TRIANGLE: MEDIAN

The length of a median drawn to a side of length c of a triangle, if the other two sides are of length a and b, is ________?

2.36

$$\frac{1}{2}\sqrt{2a^2 + 2b^2 - c^2}$$

2.37 TRIANGLE: MEDIAN

A triangular lot has sides 180 ft., 200 ft. and 240 feet long respectively. What is the length of the median drawn to the longest side of the Δ?

2.37

$$L = \frac{1}{2}\sqrt{2a^2 + 2b^2 - c^2} = \frac{1}{2}\sqrt{2 \cdot 180^2 + 2 \cdot 200^2 - 240^2}$$

$$= \frac{1}{2}\sqrt{87200} = 147.65 \text{ ft.}$$

2.38 TRIANGLE: ANGLE BISECTOR

The length of an angle bisector of a triangle, if bisecting angle A, is ________?

2.38

If the side opposite angle A is of length a, the other two sides are of length b and c, and $s = \frac{1}{2}(a + b + c)$, then the bisector length is

$$\frac{2}{b+c}\sqrt{bcs(s-a)}$$

.39

TRIANGLE: ANGLE BISECTOR

A triangular lot has sides 180 feet, 200 feet and 240 feet respectively. What is the length of the angle bisector for the largest angle of the triangle?

2.39

$$s=\frac{1}{2}(180+200+240)=310$$

$$L=\frac{2}{180+200}\sqrt{180\cdot 200\cdot 310\,(310-240)}=\frac{2}{380}\sqrt{180\cdot 200\cdot 310\cdot 70}$$

$$=\frac{2\cdot 100}{380}\sqrt{18\cdot 2\cdot 31\cdot 70}=\frac{2\cdot 100\cdot 6}{380}\sqrt{31\cdot 70}=147.11\text{ feet}$$

2.40 TRIANGLE: ALTITUDE

The length of an altitude of a triangle, if drawn to a side of length c, is ________?

2.40

If $s = \frac{a+b+c}{2}$, the altitude length is $\frac{2}{c}\sqrt{s(s-a)(s-b)(s-c)}$.

TRIANGLE

A triangular lot has sides 180 feet, 200 feet and 240 feet long respectively. What is the shortest distance from the longest side to the opposite vertex?

2.41

The shortest distance is measured by the altitude.

$$s = \frac{180+200+240}{2} = 310$$

$$L = \frac{2}{240}\sqrt{(310)(130)(110)(70)} = \frac{2 \cdot 100}{240}\sqrt{31 \cdot 13 \cdot 11 \cdot 7} = 146.8 \text{ feet.}$$

What is the locus of points equidistant from the sides of a given angle?

2.42

The bisector of the angle.

What is the locus of points equidistant from the extremities of a line?

2.43

The perpendicular bisector of the line.

In a right triangle, the altitude on the hypotenuse is the ________ ________ between the segments of the hypotenuse.

2.44

Mean proportional.

A line parallel to one side of a triangle divides the other two sides ________.

2.45

Proportionally

2.46 QUADRILATERAL

The area of any quadrilateral with diagonals a and b and the angle between them θ is ________?

2.46

$$\frac{1}{2}ab\sin\theta$$

3.1 CIRCLE

How many points in a plane, not on the same straight line, are needed to determine a unique circle?

3.1

Three

3.2 CIRCLE: ANGLE MEASURE

What is the measure of a central angle (an angle formed by two radii) of a circle?

3.2

A central angle is measured by the intercepted arc (arc intercepted by the two radii).

3.3 CIRCLE: EQUAL ARCS

What determines the equality of arcs of the same circle?

3.3

Arcs of the same circle are equal if either
1) central angles are equal, or
2) chords of these arcs are equal.

CIRCLE: CHORD

Two chords which bisect each other are ________ of the circle?

3.4

Diameters

3.5 CIRCLE: CHORD

A radius perpendicular to a chord ________ the chord and its arc.

3.5

Bisects

3.6 CIRCLE: CHORDS

What determines the equality of two chords of the same circle?

3.6

Chords of the same circle are equal if either:

1) they have equal arcs, or

2) they are equally distant from the center.

A radius drawn to the point of contact of a circle and its tangent is ________ to the tangent?

3.7

Perpendicular.

3.8 CIRCLE: ANGLE MEASURE

How is an inscribed angle of a circle measured?

3.8

An inscribed angle is measured by half its intercepted arc.

3.9 CIRCLE: ANGLE MEASURE

What is the measure of an angle inscribed in a semicircle?

3.9

A right angle or 90°.

3.10 CIRCLE: ANGLE MEASURE

What is the measure of an angle formed by a tangent and a chord of a circle?

3.10

Half of the intercepted arc.

3.11 CIRCLE: ANGLE MEASURE

What is the measure of an angle formed by two chords intersecting inside a circle?

3.11

Half the sum of the opposite intercepted arcs.

3.12 CIRCLE: TANGENT

The tangents to a circle drawn from an external point are ________?

3.12

Equal

3.13 CIRCLE: TANGENTS

The tangents to a circle drawn from an external point make ________ angles with the line joining the point to the ________ of the circle.

3.13

Equal; Center

3.14 CIRCLE: ANGLE MEASURE

What is the measure of the angle formed by two secants to a circle?

3.14

Half the difference of the intercepted arcs.

3.15 CIRCLE: ANGLE MEASURE

What is the measure of the angle formed by two tangents to a circle?

3.15

Half the difference of the intercepted arcs.

3.16 CIRCLE: LOCUS

What is the locus definition of a circle?

3.16

A circle is the locus of a point which moves so that it always remains equidistant from a fixed point.

3.17 CIRCLE: CIRCUMFERENCE

The circumference of a circle of radius r is ________?

3.17

$$2\pi r$$

3.18

CIRCLE: AREA

The area of a circle of radius r is ______?

3.18

$$\pi r^2$$

3.19 CIRCLE: CIRCUMFERENCE

Find the circumference of a circle whose area is 169π.

3.19

Circumfrence $= 2\pi r$;   Area $= \pi r^2$

$\therefore r^2 = \frac{169\pi}{\pi} = 169$; $r = 13$

Circumference $= 26\pi$.

3.20 CIRCLE

A circle is cut from a square piece of tin 20 in. on a side so as to give the least amount of waste. What amount of tin is wasted?

3.20

Area of square = 400 sq. in.
Area of inscribed circle = 100π sq. in.
Wasted tin = $(400 - 100\pi)$ sq. in.
= (400 - 314.16) sq. in.
= 85.84 sq. in.

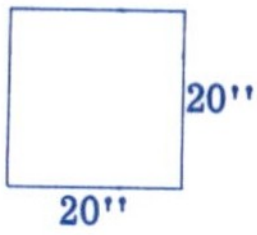

3.21 CIRCLE: CHORDS

Two parallel chords of a circle, on the same side of the center, are 6 in. and 8 in. long and 1 in. apart. Find the radius of the circle.

3.21

By the Pythagorean Theorem:

$3^2 + (1+x)^2 = r^2$

$4^2 + \quad x^2 \cdot \; = r^2$

AD = 4

BC = 3

CD = 1

OD = x

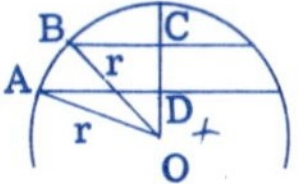

Subtracting, $7 - (1+2x) = 0$; $x = 3$

$\therefore r = \sqrt{4^2 + 3^2} = 5$ in.

3.22 CIRCLE: SECTOR

What is a sector of a circle?

3.22

A sector of a circle is a plane figure bounded by two radii and their intercepted arc.

3.23

CIRCLE: SECTOR

What is the area of a sector of a circle?

3.23

$$\text{Area} = \frac{(\text{length of arc})(\text{radius})}{2}$$

$$= \frac{\pi(\text{radius})^2(\text{angle between radii in degrees})}{360}$$

$$= \frac{(\text{radius})^2(\text{angle between radii in radians})}{2} = \frac{1}{2}r^2\theta.$$

3.24

CIRCLE: SEGMENT

What is a segment of a circle?

3.24

A segment of a circle is a plane figure bounded by a chord and its corresponding arc.

3.25 CIRCLE: SEGMENT

What is the area of a segment of a circle of radius r?

3.25

$$\text{Area} = \frac{\pi(r)^2\theta}{360} - \frac{r^2\sin\theta}{2}$$ (where θ is the angle between the two radii in degrees.)

$$= \frac{r^2}{2}(\theta - \sin\theta)$$ (where θ is in radians.)

3.26 CIRCLE: AREA

What is the area of the ring between two circles of radii r_1 and r_2, one of which encloses the other?

3.26

$$\text{Area} = \pi(r_1^2 - r_2^2) = \pi(r_1 - r_2)(r_1 + r_2)$$

Note: r_1 is the larger radius. The two circles are not necessarily concentric.

3.27

CIRCLE: AREA

A bench 2 feet wide surrounds a statue with a circular base. The area of the bench is 24π sq. ft. What is the diameter of the base of the statue?

3.27 Let r_2 = radius of base of statue.

$r_1 = r_2 + 2$ = outer base radius.

$$\text{Area of ring} = \pi(r_1 - r_2)(r_1 + r_2)$$

$$24\pi = \pi(2)(2r_2 + 2)$$

$$6 = r_2 + 1$$

$$r_2 = 5$$

$\therefore$ Base diameter = 10 ft.

CIRCLE: ARC

What is the measure of an arc of a circle of radius r?

3.28

For an arc intercepting a central angle θ, the length of the arc is equal to

1) $\frac{\pi r\theta}{180}$ (θ in degrees)

2) $r\theta$ (θ in radians)

3.29 CIRCLE: SECTOR OF ANNULUS

What is a sector of an annulus of a circle?

3.29

If a ring is formed between two concentric circles, a sector of an annulus is that part of the ring included between two radii.

3.30 CIRCLE: SECTOR OF ANNULUS

What is the area of the sector of an annulus of a circle?

3.30

If the two concentric circles have radii r_1 and r_2 ($r_1 > r_2$), and the angle between the radii is θ (in radians), the area is $\frac{1}{2}\theta(r_1+r_2)(r_1-r_2)$.

3.31 CIRCLE: PROPERTIES

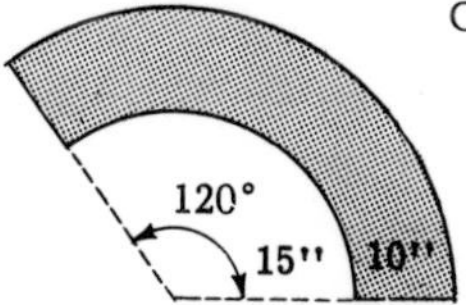

Find the area of the shaded portion.

3.31

$$\text{Area} = \frac{1}{2}\theta(r_1 + r_2)(r_1 - r_2) = \frac{1}{2}\cdot\frac{2\pi}{3}(25 + 15)(25 - 15)$$

$$= \frac{2\pi}{2\cdot 3}(40)(10) = \frac{400\pi}{3} = 418.9 \text{ sq.in.}$$

A perpendicular to a diameter from any point on the circle is a ________ ________ between the segments of the diameter.

3.32

Mean proportional.

CIRCLE: CHORDS

If two chords of a circle intersect, what can be said about the segments of those chords?

3.33

The product of the segments of one equals the product of the segments of the other.

3.34 CIRCLE: TANGENT AND SECANT

If a tangent and a secant are drawn from a point outside a circle, the tangent is the ________ ________ between the secant and its external segment.

3.34

Mean proportional.

3.35 CIRCLE: CIRCUMFERENCE

The circumference of two circles are in the same ratio as their ________?

3.35

Radii

CIRCLE: AREA

The areas of two circles are in the same ratio as the squares of their ________?

3.36

Radii, diameters, or circumferences.

4.1 POLYGON: ANGLES

The sum of the angles of a polygon of n sides is _______?

4.1

$$(n-2)\ 180^\circ\ .$$

4.2 POLYGON: ANGLES

Each angle of a regular polygon of n sides equals ______?

4.2

$$\frac{n-2}{n}(180°)$$

4.3

POLYGONS: SIMILAR

The areas of similar polygons are to each other as the ________ of corresponding sides.

4.3

Squares

4.4 POLYGON: CENTER

What is the center of a regular polygon?

4.4

The center of a regular polygon is the common center of its inscribed and circumscribed circles.

4.5 POLYGON: AREA

The area of a regular polygon is ________?

4.5

Half the product of its perimeter and the radius of its inscribed circle.

4.6 POLYGONS: SIMILAR

What determines similar polygons?

4.6

Two polygons are similar if their corresponding angles are equal and their corresponding sides are proportional.

4.7 POLYGON: HEXAGON

The area of a regular hexagon of side s is________.

4.7

$$\frac{3\sqrt{3}}{2}\,s^2$$

4.8 POLYGON: REGULAR

The area of a regular polygon of n sides, each of length s, is ________.

4.8

$$\frac{1}{4} ns^2 \cot \frac{180°}{n}$$

4.9 POLYGON: REGULAR

The radius of the inscribed circle for a regular polygon of n sides each of length s, is ________.

4.9

$$\frac{s}{2}\cot\frac{180^\circ}{n}$$

4.10 POLYGON: CIRCUMSCRIBED CIRCLE

The radius of the circumscribed circle for a regular polygon of n sides each of length s is ________.

4.10

$$\frac{s}{2} \csc \frac{180°}{n}$$

4.11 POLYGONS

What are the names of polygons having 5, 6, 8 or 10 sides?

4.11

5 sides - pentagon
6 sides - hexagon
8 sides - octagon
10 sides - decagon.

4.12 POLYGON: SIMPSON'S RULE

What is Simpson's rule for finding the approximate area of an irregular plane figure?

4.12 $A = \frac{h}{3}(y_0 + 4y_1 + 2y_2 + \ldots$

$+ 2y_{n-2} + 4y_{n-1} + y_n)$

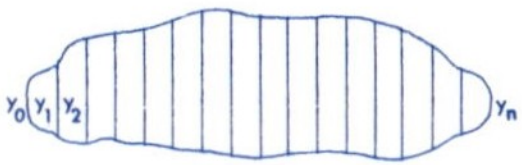

Divide the figure into an even number of strips or panels by a series of equidistant parallel chords y_0, y_1, y_2, ----, y_n, the common distance between the chords being h. The required approximate area is then given above.

5.1 PLANE

What is a plane?

5.1

A plane is a surface that wholly contains a straight line that connects any two points in that surface.

5.2 PLANE

How many unique points in space, not all on the same line, determine a unique plane?

5.2

Three

5.3 LINE AND PLANE: PERPENDICULAR

What determines perpendicularity of a line and a plane?

5.3

A line is perpendicular to a plane if the line is perpendicular to every straight line drawn through its foot in the plane.

5.4 LINE AND PLANE: PROJECTION

What is meant by the projection of a point on a plane?

5.4

The projection of a point on a plane is the foot of the perpendicular drawn from the point to the plane.

5.5 LINE AND PLANE: PROJECTION

What is meant by the projection of a line on a plane?

5.5

The projection of a line on a plane is the locus of the projection of all its points.

5.6 LINES AND PLANES: ANGLE

What is meant by the angle which a line makes with a plane?

5.6

The angle which a line makes with a plane is the angle which it makes with its projection on the plane.

5.7

PLANE

What determines a plane?

5.7

A plane is determined by: 1) a straight line and a point not on the line. 2) two intersecting straight lines. 3) two parallel lines. 4) three points not on the same straight line.

5.8 LINES IN SPACE

What determines parallel lines in space?

5.8

1) Intersection of two parallel planes with a third plane. 2) Two lines perpendicular to the same plane. 3) Two lines parallel to the same line.

5.9 PLANES: PARALLEL

What determines parallel planes?

5.9

1) If two intersecting lines are parallel to a plane, their plane is parallel to that plane 2) two planes perpendicular to the same line are parallel.

5.10 PLANE: DIHEDRAL ANGLE

What is a dihedral angle?

5.10

A dihedral angle is the amount of divergence of two intersecting planes.

5.11 PLANES: DIHEDRAL ANGLE

What is the measure of dihedral angle?

5.11

If two lines are formed, one in each plane, and both perpendicular to the edge at the same point, the angle formed by these two lines is called the plane angle. The plane angle is taken as the measure of the dihedral angle.

What is the locus of points equidistant from two given points?

5.12

A plane that is the perpendicular bisector of the line segment joining them.

5.13 LINES IN SPACE: SKEW

What are skew lines?

5.13

Two lines in space that are neither parallel nor intersecting are called skew lines.

5.14 LINES IN SPACE: SKEW

How many perpendiculars can be drawn between two skew lines?

5.14

One and only one.

5.15 POLYHEDRAL ANGLE

How is a polyhedral angle formed?

5.15

A polyhedral angle is formed when three or more distinct planes intersect at a point.

5.16 POLYHEDRAL ANGLE

The sum of the face angles of a polyhedral angle have to be less than _______?

5.16

The sum is less than 360°.

6.1 POLYHEDRON: SOLID

What is a solid?

6.1

A solid is a limited portion of space, bounded by surfaces.

6.2 POLYHEDRON: SECTION

What is a section of a solid?

6.2

A section of a solid is the plane figure cut from the solid by passing a plane through it.

6.3 POLYHEDRON

What is a polyhedron?

6.3

A polyhedron is a solid bounded by planes.

6.4 POLYHEDRON: EDGES

What are the edges of a polyhedron?

6.4

The edges are the intersections of the bounding planes.

6.5 POLYHEDRON: FACES

What are the faces of a polyhedron?

6.5

The faces are polygons, which are the portions of the bounding planes included by the edges.

6.6 POLYHEDRON: VERTICES

What are the vertices of a polyhedron?

6.6

The vertices are the intersections of the edges.

POLYHEDRONS

What is a regular polyhedron?

6.7

A regular polyhedron is one whose faces are all congruent regular polygons.

6.8 POLYHEDRONS

What are the names of the only five regular polyhedrons which exist?

6.8

tetrahedron -- one with four faces
hexahedron (cube) -- one with six faces
octahedron -- one with eight faces
dodecahedron -- one with twelve faces
icosahedron -- one with twenty faces

6.9 POLYHEDRONS: EDGES

What are the number of edges on each of the regular polyhedrons?

6.9

6, 12, 12, 30, 30

6.10 POLYHEDRONS: VERTICES

What are the number of vertices on each of the regular polyhedrons?

6.10

4, 8, 6, 20, 12

6.11 POLYHEDRON: CUBE

What is a cube?

6.11

A cube is a polyhedron whose six faces are all squares.

6.12 POLYHEDRON: CUBE

What is the total area of a cube of edge a?

6.12

$6a^2$

6.13 POLYHEDRON: CUBE

What is the volume of a cube of edge a?

6.13

a^3

6.14 POLYHEDRON: CUBE

A cube of edge a is cut by a plane containing two diagonally opposite edges of the cube. What is the area of the section thus formed?

6.14

A rectangle is formed, one side of which is equal to a, the other equal to the diagonal of a face $(a\sqrt{2})$ Therefore the area is $a(a\sqrt{2}) = \sqrt{2}\,a^2$

6.15 POLYHEDRON: CUBE

What is the length of a diagonal of a cube of edge a?

6.15

$$(AC)^2 = (AB)^2 + (BC)^2$$

$$(BC)^2 = (BD)^2 + (DC)^2 = 2a^2$$

$$\therefore (AC)^2 = a^2 + 2a^2 = 3a^2$$

$$AC = a\sqrt{3}$$

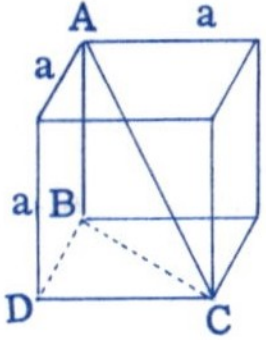

6.16 POLYHEDRON: CUBE

Express the total area and the volume of a cube in terms of its diagonal d.

6.16

If the cube has edge a, the diagonal has a length of $a\sqrt{3}$.

$$d = a\sqrt{3}; \quad \text{Area} = 6a^2 = 6\frac{d^2}{3} = 2d^2$$

$$a = \frac{d}{\sqrt{3}}; \quad \text{Volume} = a^3 = \frac{d^3}{3\sqrt{3}} = \frac{d^3\sqrt{3}}{9}$$

6.17 POLYHEDRON: CUBE

Find the edge of a cube whose volume and entire surface contain the same number of units.

6.17

Let a be the edge.

Then $6a^2 = a^3$; $a^3 - 6a^2 = 0$; $a^2(a-6) = 0$

$a = 6$ units.

6.18 POLYHEDRON: PARALLELEPIPED

What is a rectangular parallelepiped?

6.18

A rectangular parallelepiped is a polyhedron whose six faces are all rectangles.

6.19 POLYHEDRON: PARALLELEPIPED

What is the total area of a rectangular parallelepiped of edges a, b and c?

6.19

$$2\ (ab + bc + ca)$$

6.20 POLYHEDRON: PARALLELEPIPED

What is the volume of a rectangular parallelepiped of edges a, b and c?

6.20

abc

6.21 POLYHEDRON: PARALLELEPIPED

What is the length of the diagonal of a rectangular parallelepiped of edges a, b, c?

6.21

$$(AB)^2 = (AC)^2 + (CB)^2$$

$$(CB)^2 = c^2 + a^2$$

$$\therefore (AB)^2 = b^2 + c^2 + a^2$$

$$AB = \sqrt{a^2 + b^2 + c^2}$$

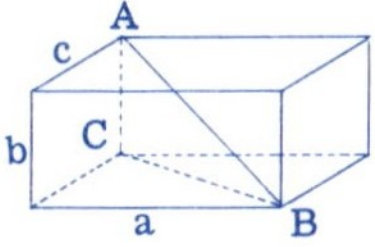

6.22 POLYHEDRON: PARALLELEPIPED

How many square feet of material are needed to line the bottom and sides of a rectangular cistern 10 ft. long, 7 ft. wide and 4.5 ft. deep?

6.22

$$\text{Area} = (\text{length} \times \text{width}) + 2\,(\text{length} \times \text{depth}) + 2\,(\text{width} \times \text{depth}) = (10 \times 7) + 2\,(10 \times 4.5) + 2\,(7 \times 4.5)$$

$$= 70 + 90 + 63 = 223 \text{ sq. ft.}$$

6.23 POLYHEDRON: PARALLELEPIPED

How much dirt is removed in excavating a basement 40 ft. by 30 ft. by 7 ft.?

6.23

$$\text{Volume} = 40 \times 30 \times 7$$
$$= 8400 \text{ cu. ft.}$$

6.24 POLYHEDRON: PRISM

What is a prism?

6.24

A prism is a polyhedron of which two faces are equal polygons (bases) in parallel planes, and the other faces are parallelograms.

6.25 POLYHEDRON: LATERAL AREA

What is meant by lateral area?

6.25

Lateral area is the sum of the areas of the faces, excluding the area of the bases.

6.26 POLYHEDRON: PRISM

What is the altitude of the prism?

6.26

The altitude is the perpendicular distance between the planes of the bases.

6.27 POLYHEDRON: PRISM

What is a right section of a prism?

6.27

A right section is a section perpendicular to the lateral edges.

6.28 POLYHEDRON

A dam 40 feet high has a uniform cross section as shown. The dam is 80 feet long. Find the volume of material in the dam.

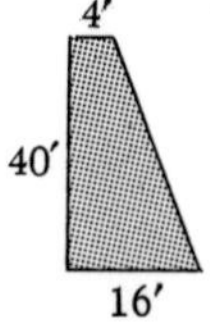

6.28

The dam is a prism of altitude 80 feet and with a trapezoidal base.

Base area $= \frac{1}{2}(4+16)(40) = 400$ sq. ft.

Volume = (Base)(altitude) = (400)(80) = 32,000 cu.ft.

6.29 POLYHEDRON: PRISM

Find the volume of a triangular prism with the base having sides 3, 4 and 5; and altitude 6.

6.29

The base is a right triangle, as

$5^2 = 3^2 + 4^2$. Base area $= \frac{1}{2}(3)(4) = 6$.

Volume = (Base)(Altitude) = (6)(6) = 36.

6.30 POLYHEDRON: PRISM

What is a right prism?

6.30

A right prism is a prism whose lateral edges are perpendicular to its bases; its lateral faces are rectangles.

6.31 POLYHEDRON: PRISM

What is the lateral area of a prism?

6.31

Lateral area = (length of lateral edge) · (perimeter of the right section).

POLYHEDRON: PRISM

What is the volume of a prism?

6.32

The volume of a prism is either

1) (base)·(altitude)

or

2) (right section)·(lateral edge)

6.33 POLYHEDRON: PRISM

Find the lateral area and total area of a right triangular prism with the sides of base 8, 9, 10 and altitude 6.

6.33

Lateral area $= 6(8+9+10) = (6)(27) = 162.$

Total area = Lateral area + Base area

Base area $= \sqrt{s(s-a)(s-b)(s-c)} = \sqrt{(13.5)(5.5)(4.5)(3.5)} = 34.2$

Total area $= 162 + 34.2 = 196.2.$

The volume of a prism is 540 cu. in., its altitude is 10 in. and its base is a regular hexagon. Find the side of the base.

6.34

Volume = (Base area)(Altitude)

540 = (Base area)·10

∴ Base area = 54

If s is the side of the hexagon, the area is $\frac{3\sqrt{3}}{2}s^2$.

$$\frac{3\sqrt{3}\,s^2}{2} = 54; \quad s^2 = \frac{108}{3\sqrt{3}} = 12\sqrt{3}; \quad s = 4.56 \text{ in.}$$

6.35 POLYHEDRON: PYRAMID

What is a pyramid?

6.35

A pyramid is a polyhedron of which one face (called the base) is a polygon of any number of sides; and the other faces are triangles which have a common vertex (called the lateral faces).

6.36 POLYHEDRON: PYRAMID

What is the altitude of a pyramid?

6.36

The altitude is the length of the perpendicular dropped from the vertex to the plane of the base.

6.37 POLYHEDRON: PYRAMID

The ratio of parallel sections of a pyramid are equal to ________?

6.37

The ratios of the squares of the corresponding distances from the sections to the vertex of the pyramid.

6.38

POLYHEDRON: PYRAMID

If a pyramid is cut by a plane parallel to the base, what can be said about the lateral edges and the altitude?

6.38

They are divided proportionally.

6.39

POLYHEDRON: PYRAMID

What is the lateral area of a pyramid?

6.39

The lateral area is equal to the sum of areas of faces.

6.40

PYRAMID: VOLUME

What is the volume of a pyramid?

6.40

The volume of any pyramid is equal to one-third the product of the base and the altitude.

6.41 PYRAMID: REGULAR

What is a regular pyramid?

6.41

A regular pyramid is one whose base is a regular polygon whose center coincides with the foot of the perpendicular dropped from the vertex to the base.

6.42 POLYHEDRON: PYRAMID

The lateral faces of a regular pyramid are ________.

6.42

Congruent isosceles triangles.

6.43 POLYHEDRON: PYRAMID

What is the slant height of a regular pyramid?

6.43

The slant height of a regular pyramid is the altitude of a lateral face.

POLYHEDRON: PYRAMID

What is the altitude of a regular pyramid?

6.44

The altitude of a regular pyramid is equal to the length of the perpendicular dropped from the vertex to the center of the base.

6.45 POLYHEDRON: PYRAMID

What is the lateral area of a regular pyramid?

6.45

Lateral area = $\frac{1}{2}$ (perimeter of base) . (slant height)

6.46

POLYHEDRON: PYRAMID

What is the volume of a regular pyramid?

6.46

The volume is equal to one-third the product of the base and altitude.

6.47 POLYHEDRON: PYRAMID

Find the volume of a regular pyramid with a square base 7 ft. on a side and height 9 feet.

6.47

$$\text{Volume} = \frac{1}{3}(\text{Base})(\text{Altitude}) = \frac{1}{3}(49)(9) = (49)(3) = 147 \text{ cu. ft.}$$

POLYHEDRON: PYRAMID

What is a frustum of a regular pyramid?

6.48

A frustum of a regular pyramid is the portion of a regular pyramid included between the base and a section parallel to the base.

6.49 POLYHEDRON: PYRAMID

The faces of a frustum of a regular pyramid are ______?

6.49

Congruent isosceles trapezoids.

6.50

POLYHEDRON: PYRAMID

What is the lateral area of the frustum of a regular pyramid?

6.50

$$\text{Lateral area} = \frac{(\text{sum of perimeters of bases})}{2} \cdot (\text{slant ht.})$$

$$= (\text{mean perimeter}) \cdot (\text{slant height})$$

6.51

POLYHEDRON: PYRAMID

What is the volume of the frustum of any pyramid?

6.51

If the two base areas are B_1 and B_2, and the altitude is h,

$$\text{the volume} = \frac{(B_1 + B_2 + \sqrt{B_1 B_2})}{3} \cdot h$$

POLYHEDRON: PYRAMID

6.52

Find the volume of the frustum of a pyramid with base areas $B_1 = 8$ sq. ft., $B_2 = 2$ sq. ft. and altitude $h = 3$ ft.

6.52

$$V = \frac{(B_1 + B_2 + \sqrt{B_1 B_2})}{3} \cdot h = \frac{8+2+\sqrt{16}}{3} \cdot 3 = \frac{14}{3} \cdot 3 = 14 \text{ cu. ft.}$$

6.53

A pyramid 4 ft. high with a base area of 8 sq. ft. is cut by a plane parallel to the base and 2 ft. from it. Find the volume of the frustum.

6.53 $V_1 = \text{Volume (original pyramid)} = \frac{1}{3} \cdot 8 \cdot 4 = \frac{32}{3}$ cu. ft.

$V_2 = \text{Volume (pyramid removed)} = \frac{1}{3}(\text{Base}) \cdot 2$

But $\frac{B_1}{B_2} = \frac{h_1^2}{h_2^2} = \frac{4^2}{2^2} = \frac{4}{1}$.

$\therefore B_2 = \frac{1}{4}(8) = 2$ sq. ft. $\qquad \therefore V_2 = \frac{1}{3}(2)(2) = \frac{4}{3}$ cu. ft.

Volume of frustum $= \frac{32}{3} - \frac{4}{3} = \frac{28}{3}$ cu. ft.

6.54

POLYHEDRONS: SIMILAR

What is meant by similar polyhedrons?

6.54

Similar polyhedrons having the same number of faces, respectively similar and similarly placed, and having their corresponding polyhedral angles equal.

6.55

POLYHEDRON: SIMILAR FIGURES

The areas of similar plane figures on surfaces have the same ratio as ________ of any two corresponding lines.

6.55

The squares.

6.56

POLYHEDRONS: SIMILAR

The volumes of similar solids have the same ratio as the ________ of any two corresponding lines.

6.56

Cubes

6.57

POLYHEDRON: PRISMATOID

What is a prismatoid?

6.57

A prismatoid is a polyhedron having for bases two polygons in parallel planes; and for lateral faces, triangles, trapezpoids or parallelograms with one side lying in one base, and the opposite side or vertex lying in the other base.

6.58

POLYHEDRON: PRISMATOID

What is the mid-section of a prismatoid?

6.58

A section parallel to the bases and midway between them.

6.59 POLYHEDRON: PRISMATOID

What is the volume of a prismatoid having bases of area B_1 and B_2, a midsection of area M, and altitude h?

6.59

$$\frac{1}{6}\left[B_1 + B_2 + 4M\right] \cdot h$$

This is called the prismoidal formula.

6.60

PRISMOIDAL FORMULA

Where can the prismoidal formula be used, either to give exact volume or approximations?

6.60

The prismoidal formula gives the volumes exactly for a cone, sphere, ellipsoid and prismoid, and can be used as a good approximation for many other solids.

6.61 PRISMOIDAL FORMULA

Use the prismoidal formula to find the volume of the truncated triangular prism shown.

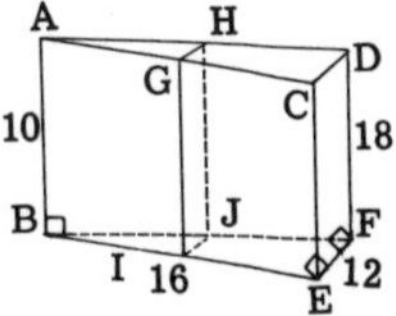

6.61

B_1 is line AB, B_2 area of face CDEF. In midsection GHIJ: $IG = \frac{18+10}{2} = 14$; $GH = \frac{12}{2} = 6$

$\therefore B_1 = 0, \quad B_2 = 18 \cdot 12 = 216, \quad M = 14 \cdot 6 = 84$

$$V = \frac{1}{6}(16)(0+216+4 \cdot 84) = 1472$$

6.62 POLYHEDRON

What is Cavalieri's Theorem?

6.62

If in two solids of equal heights the sections made by planes parallel to and at the same distance from their respective plane bases are always equal, the volumes of the solids are equal.

7.1 CYLINDER: SURFACE

What is a cylindrical surface?

7.1

A cylindrical surface is a surface generated by a movin straight line, called the generator, which is always para llel to a fixed line; and which always intersects a fixe plane curve, called the directrix, not in the plane witl the fixed line.

7.2 CYLINDER: ELEMENT

What is an element of a cylindrical surface?

7.2

An element is a generator in any particular position.

7.3 CYLINDER

What is a cylinder?

7.3

A cylinder is a solid bounded by a closed cylindrical surface and two parallel planes. The bounding cylindrical surface is called the lateral surface; the bounding parallel planes, the bases.

7.4 CYLINDER: RIGHT SECTION

What is a right section of a cylinder?

7.4

A right section is a section perpendicular to all elements of the cylinder.

7.5 CYLINDER: LATERAL AREA

What is the lateral area of a cylinder?

7.5

Lateral area = (perimeter of right section) (length of element).

7.6 CYLINDER: VOLUME

What is the volume of a cylinder?

7.6

Volume =

1) (base) · (altitude)

2) (right section) · (element)

7.7 CYLINDER: CIRCULAR

What is a circular cylinder?

7.7

A circular cylinder is a cylinder which has a circular right section.

7.8 CYLINDER: AREA AND VOLUME

What is the lateral area and volume of a circular cylinder?

7.8

The same as for any cylinder.

i.e. base x altitude

7.9 CYLINDERS: SIMILAR

What are similar circular cylinders?

7.9

Similar circular cylinders are obtained if the corresponding elements of the surface and the radii of the bases are proportional.

7.10 CYLINDER: AREA

The ratio of the lateral areas, or total areas, of two similar cylinders is equal to ________?

7.10

The ratios of the squares of the corresponding dimensions (radii, altitudes or elements).

7.11 CYLINDERS: SIMILAR

The ratio of the volumes of two similar circular cylinders is equal to ________?

7.11

The ratios of the cubes of the corresponding dimensions (radii, altitudes or elements).

7.12 CYLINDER: RIGHT CIRCULAR

What is a right circular cylinder?

7.12

A right circular cylinder is a circular cylinder whose elements are perpendicular to its base. It is also called a cylinder of revolution, as it can be generated by the revolution of a rectangle about one side as an axis.

7.13 CYLINDER: AXIS

What is the axis of a right circular cylinder?

7.13

The axis is the line joining the centers of the base.

7.14 CYLINDER

What is the lateral area of a right circular cylinder of base radius r and altitude h?

7.14

$$2\pi rh$$

7.15 CYLINDER: VOLUME

What is the volume of a right circular cylinder of base radius r and altitude h?

7.15

$$\pi r^2 h$$

7.16 CYLINDER - SURFACE AND VOLUME

Find the total surface and volume of a right circular cylinder with radius 7 feet and height 7 feet.

7.16

Total surface = Lateral area + Base area

$$= 2\pi rh + 2\pi r^2 = 98\pi + 98\pi = 196\pi = 616 \text{ sq. ft.}$$

Volume $= \pi r^2 h = 343\pi = 1078$ cu. ft.

7.17 CYLINDERS

If two similar circular cylinders have radii of 10 inches and 12 inches, find the ratio of their volumes.

7.17

$$\frac{V_1}{V_2} = \frac{r_1^{\,3}}{r_2^{\,3}}$$

$$\frac{V_1}{V_2} = \frac{10^3}{12^3} = \frac{5^3}{6^3} = \frac{125}{216}$$

8.1 CONE: SURFACE

What is a conical surface?

8.1

A conical surface is a surface generated by a moving straight line (generator) which always intersects a fixed plane curve (called the directrix) and which always passes through a fixed point (called the vertex) not in the plane of the curve.

8.2 CONE: TANGENT PLANE

When is a plane tangent to a conical surface?

8.2

A plane is tangent to a conical surface if it contains an element of the conical surface and a line tangent to the surface.

8.3 CONE

What is a cone?

8.3

A cone is the solid bounded by a conical surface (the lateral surface) whose directrix is a closed curve, and a plane (the base) which cuts all the elements.

8.4 CONE: ALTITUDE

What is the altitude of a cone?

8.4

The altitude is the perpendicular distance from the vertex to the plane of the base.

8.5 CONE: SECTION

The section of a cone made by a plane passing through its vertex and containing two points of the base is a ________?

8.5

Triangle

8.6 CONE: AXIS

What is the axis of a cone?

8.6

The axis of a cone is the straight line joining the vertex to the center of the base (if the base has a center).

8.7 CONE: RIGHT SECTION

What is the right section of a cone?

8.7

A right section of a cone is a section perpendicular to its axis and cutting all the elements.

8.8 CONE: CIRCULAR

What is a circular cone?

8.8

A circular cone is a cone whose right section is a circle.

8.9

CONE: VOLUME

What is the volume of a cone?

8.9

The volume of a cone = $(\frac{1}{3}$ base area$)\cdot($altitude$)$.

8.10 CONE: RIGHT CIRCULAR

What is a right circular cone?

8.10

A right circular cone is a circular cone whose axis is perpendicular to its base.

8.11 CONE: SLANT HEIGHT

What is the slant height of a circular cone?

8.11

The slant height is the length of an element.

8.12 CONE: SECTION

A section of a right circular cone which contains the vertex and two points of the base is an _______?

8.12

Isosceles triangle.

8.13 CONE: LATERAL AREA

What is the lateral area of a right circular cone?

8.13

$$\text{Lateral area} = \frac{1}{2}(\text{Circumference of base})\cdot(\text{Slant height})$$

$$= \frac{1}{2}(2\pi r)\sqrt{r^2 + h^2} = \pi r\sqrt{r^2 + h^2}$$

8.14 CONE: VOLUME

What is the volume of a right circular cone of base radius r and altitude h?

8.14

$$\frac{1}{3}\pi r^2 h$$

8.15 CONE: RIGHT CIRCULAR

If a right circular cone has a base area equal to 50π and a volume of 500π, find the base radius r and the altitude h.

8.15

$$\text{Area} = \pi r^2 = 50\pi \qquad \text{Volume} = \frac{1}{3}\pi r^2 h = 500\pi$$

$$r^2 = 50 \qquad r^2 h = 1500$$

$$r = 5\sqrt{2}$$

$$h = \frac{1500}{r^2} = \frac{1500}{50} = 30$$

8.16 CONE: AREA

How many square yards of canvas will be required to make a conical tent 15 feet high and 18 feet in diameter if 10% of the material is wasted?

8.16

$$\text{Lateral area} = \pi r\sqrt{r^2+h^2} = \pi 9\sqrt{9^2+15^2}\ \text{sq. ft.}$$

$$= \pi\sqrt{9^2+15^2}\ \text{sq. yd.} = 17.5\pi\ \text{sq. yd.}$$

$$\text{Total material} = \frac{(17.5)\pi}{.90} = 61\ \text{sq. yd.}$$

8.17 CONE: FRUSTUM

What is the frustum of a right circular cone?

8.17

The frustum is that portion of a right circular cone included between the base and a section parallel to the base.

8.18 CONE: LATERAL AREA

What is the lateral area of the frustum of a right circular cone?

8.18

Lateral area =

$$\frac{1}{2}(\text{sum of circumferences of bases})\cdot(\text{slant height})$$

$$= (\text{Mean Circumference})\cdot(\text{Slant height})$$

8.19 CONE: FRUSTUM LATERAL AREA

What is an alternative form for the lateral area of the frustum of a right circular cone, radii of bases r_1 and r_2 and altitude h?

8.19

$$\text{Lateral area} = \theta(r_1 + r_2)\sqrt{h^2 + (r_1 - r_2)^2}$$

8.20 CONE: VOLUME OF FRUSTUM

What is the volume of the frustum of a right circular cone of base areas B_1 and B_2 and altitude h?

8.20

$$\text{Volume} = \frac{\left(B_1 + B_2 + \sqrt{B_1 B_2}\right)}{3} \cdot h$$

8.21 CONE: FRUSTUM

What is an alternative form for the volume of a frustum of a right circular cone, with base radii r_1 and r_2 and altitude h?

8.21

$$\text{Volume} = \frac{\pi h}{3}\left(r_1^2 + r_1 r_2 + r_2^2\right)$$

8.22 CONE: FRUSTUM

Find the volume of the frustum of a cone with base radii 5 ft. and 3 ft. and altitude 3 ft.

8.22

$$\text{Volume} = \frac{\pi \cdot 3}{3}(5^2 + 5\ 3 + 3^2)$$

$$= \pi(25 + 15 + 9)$$

$$= 49\pi$$

9.1 SPHERE

What is a sphere?

9.1

A sphere is a solid bounded by a closed surface, every point of which is equidistant from a fixed point called the center.

9.2 SPHERE: SECTION

Every plane section of a sphere is a ________?

9.2

Circle

9.3 SPHERE: GREAT CIRCLE

What is a great circle of a sphere?

9.3

A great circle is a plane section which contains a diameter of the sphere.

SPHERE: AXIS

What is the axis of a circle of a sphere?

9.4

The axis is the diameter of the sphere perpendicular to the plane of the circle.

9.5 SPHERE: POLES

What are the poles of a circle of the sphere?

9.5

The poles are the ends of its axis.

9.6 SPHERICAL LINES

What is the shortest line that can be drawn on the surface of a sphere between two points?

9.6

The shorter arc of the great circle passing through the points.

9.7 SPHERE: SURFACE AREA

What is the area of the surface of a sphere of radius r?

9.7

$$4\pi r^2$$

9.8 SPHERE: VOLUME

What is the volume of a sphere of radius r?

9.8

$$\frac{4}{3}\pi r^3$$

9.9 SPHERE: PROPERTIES OF

What is the surface and volume of a sphere whose diameter is 15 inches?

9.9

$$\text{Surface area} = 4\pi r^2 = 4\pi\left(\frac{d}{2}\right)^2 = \pi d^2 = 225\,\pi \text{ sq. in.}$$

$$\text{Volume} = \frac{4}{3}\pi r^3 = \frac{4}{3}\pi\left(\frac{d}{2}\right)^3 = \frac{\pi d^3}{6} = \frac{\pi \cdot 15^3}{6} = \frac{\pi \cdot 15^2 \cdot 5}{2} = \frac{1125}{2}\,\pi \text{ cu. in.}$$

9.10 SPHERES: VOLUME

The volumes of two spheres are proportional to ______.

9.10

The cubes of the corresponding radii or diameters.

9.11 SPHERE: SURFACE AREA

The surface areas of two spheres are proportional to __________.

9.11

The squares of the corresponding radii or diameters.

What is the ratio of the areas and the volume of two spheres whose radii are 8 and 12?

9.12

$$\frac{A_1}{A_2} = \frac{r_1^{\,2}}{r_2^{\,2}} = \frac{8^2}{12^2} = \frac{64}{144} = \frac{4}{9}$$

$$\frac{V_1}{V_2} = \frac{r_1^{\,3}}{r_2^{\,3}} = \frac{8^3}{12^3} = \left(\frac{8}{12}\right)^3 = \frac{8}{27}$$

9.13 SPHERES

Grapefruit 4 in. in diameter sell for 25¢/ doz. What should be the price for grapefruit 5 in. in diameter?

9.13

$$\frac{C_1}{C_2} = \frac{V_1}{V_2} = \frac{d_1{}^3}{d_2{}^3} = \frac{4^3}{5^3}$$

$$\frac{25}{C_2} = \frac{4^3}{5^3}$$

$$C_2 = 25\left(\frac{5}{4}\right)^3 = 25\,(1.25)^3 = 48.8 \text{ or } 49¢/\text{doz.}$$

9.14 CONE: LATERAL AREA

A right circular cone with radius of base $\frac{a}{2}$ is inscribed in a sphere of radius a. What is the lateral area of the cone?

9.14

A cross section is shown.

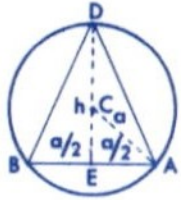

$$h = DE = DC + CE = a + \sqrt{a^2 - \left(\frac{a}{2}\right)^2} = a + \frac{a}{2}\sqrt{3}$$

$$\text{Lateral area} = \pi r\sqrt{r^2 + h^2} = \pi a\sqrt{\frac{a^2}{4} + a^2 + a^2\sqrt{3} + \frac{3}{4}a^2}$$

$$= \pi a^2\sqrt{2 + \sqrt{3}} = 5.05\, a^2$$

CYLINDER: VOLUME

What is the volume of a right circular cylinder of radius a inscribed in a sphere of radius 2a?

9.15

A cross section is shown.

$$\left(\frac{h}{2}\right)^2 = 4a^2 - a^2 = 3a^2$$

$$\frac{h}{2} = \sqrt{3}\,a$$

$$h = 2\sqrt{3}\,a$$

$$V = \pi \cdot a^2 (2\sqrt{3}\,a) = 2\sqrt{3}\,a^3 \pi$$

9.16 SPHERE: ZONE

What is a zone of a sphere?

9.16

A zone is that portion of the surface of a sphere included between two parallel planes. It is also called a spherical segment, if it is a zone of one base (one plane is tangent to the sphere).

9.17 SPHERICAL SEGMENT

What is the area of a spherical segment of a sphere of radius r if the segment has altitude h?

9.17

$$2\pi r h$$

SPHERE: ZONE

What is the volume of a zone of a sphere, if the two bases of the zone have radii r_1 and r_2 and the zone has altitude h?

9.18

$$\frac{1}{2}\pi h(r_1^2 + r_2^2) + \frac{1}{6}\pi h^3$$

9.19 SPHERICAL SEGMENT

What is the volume of a spherical segment of a sphere of height h, radius of sphere r?

9.19

$$V = \frac{1}{3}\pi h^2 (3r - h)$$

SPHERICAL SEGMENT

Find the volume of a spherical segment 4 in. in height on a sphere of radius 6 in.

9.20

$$V = \frac{1}{3}\pi \cdot 4^2(3\cdot 6 - 4) = \frac{16\pi}{3}(14) = \frac{224\pi}{3} = 234.6 \text{ cu. in.}$$

9.21 SPHERICAL ANGLE

What is a spherical angle?

9.21

An angle formed by two arcs of great circles on a sphere. It is the plane angle formed by the lines tangent to the arcs at their point of intersection.

9.22 SPHERICAL ANGLE

What is the measure of a spherical angle?

9.22

It is measured by the arc of the great circle that has the vertex of the angle as a pole and is included between the sides of the angle.

9.23 SPHERE: LUNE

What is a lune?

9.23

A lune is the portion of the surface of a sphere bounded by two semicircles of great circles.

SPHERE: LUNE

What is the area of a lune of angle θ, if the sphere has radius r?

9.24

$\frac{\pi r^2 \theta}{90}$, θ in degrees

or $2r^2 \theta$, θ in radians.

9.25 SPHERE: LUNE

What is the area of a lune whose angle is $60°$ on a sphere whose radius is 4 inches?

9.25

$$A = \frac{\pi . 4^2 . 60}{90} = \frac{32\pi}{3} \text{ sq. in.}$$

9.26 SPHERICAL POLYGON

What is a spherical polygon?

9.26

A spherical polygon is the portion of the surface of a sphere bounded by three or more arcs of great circles.

9.27 SPHERICAL POLYGON: AREA

What is the area of a spherical polygon of n sides where θ is the sum of its angles in radians and r is the radius of the sphere?

9.27

$$\text{Area} = \left[\theta - (n-2)\,\pi\right] r^2 .$$

What is the area of a spherical hexagon each of whose angles is 130°, on a sphere of radius 9?

9.28

$$\theta = \cancel{6}\cdot 13\cancel{0}\cdot\frac{\pi}{\cancel{180}_{3}} = \frac{13}{3}\pi$$

$$\text{Area} = \left[\frac{13}{3}\pi - 4\pi\right]\cdot 9^2 = \frac{\pi}{3}\cdot 81 = 27\pi$$

9.29 SPHERICAL TRIANGLE

What is a spherical triangle?

9.29

A spherical polygon of three sides.

9.30 SPHERICAL TRIANGLE

What is the area of a spherical triangle whose angles are A, B and C (in radians) on a sphere of radius r?

9.30

$$\text{Area} = (A + B + C - \pi)\, r^2$$

9.31 SPHERICAL TRIANGLE

What is meant by a: a) right, b) birectangular or c) tri-rectangular spherical triangle?

9.31

a) By a right spherical triangle is meant one which has one right spherical angle.
b) By birectangular spherical triangle is meant one which has two right spherical angles.
c) By trirectangular spherical triangle is meant one which has three right spherical angles.

9.32 SPHERICAL TRIANGLE

What is the area of a trirectangular spherical triangle on a sphere of radius 12 in.?

9.32

$$A = B = C = 90^\circ = \frac{\pi}{2} \text{ radians}$$

$$\text{Area} = \left(\frac{3\pi}{2} - \pi\right) 12^2 = 72\pi \text{ sq. in.}$$

9.33 SPHERICAL TRIANGLE

A spherical triangle has an area of 64 on a sphere whose area is 400. If two of the angles of the triangle are 107° and 118°, find the third angle.

9.33

$400 = 4\pi r^2, \quad r = \sqrt{\frac{100}{\pi}}$

$$64 = \left(\frac{107\cdot\pi}{180} + \frac{118\pi}{180} + \frac{C\pi}{180} - \pi\right)\cdot\left(\frac{100}{\pi}\right) = \left(\frac{107}{180} + \frac{118}{180} + \frac{C}{180} - 1\right) 100$$

$.64 = \frac{45}{180} + \frac{C}{180} \qquad C = 180(.64) - 45 = 70^\circ\!.2$

9.34 SPHERICAL TRIANGLE

What are the limiting values for the sum of the angles of a convex spherical triangle?

9.34

The sum is greater than 180° but less than 540°.

10.1 TRIGONOMETRY: DEFINITION

What is trigonometry?

10.1

Trigonometry is the study of certain ratios associated with an angle; properties of these ratios; their interrelationships; and their application to various problems.

10.2 TRIG. DEFINITIONS: ANGLE

How is an angle generated?

10.2

An angle is generated by rotating a line about one end, from an initial position to a terminal position. The size of the angle is determined by the amount of rotation.

10.3 ANGLE: POSITIVE OR NEGATIVE

What determines a positive or negative angle?

10.3

An angle is considered positive if the line is rotated counterclockwise from its initial position; negative if the rotation is clockwise.

10.4 ANGLE MEASURE: SYSTEMS

What are the two main systems of measuring angles?

10.4

The sexagesimal system (degrees, minutes, seconds) and the radian measure.

10.5 ANGLE MEASURE - SEXAGESIMAL

What is the sexagesimal system of measure of angles?

10.5

One revolution of the rotating line measures 360°; each degree is divided into 60 minutes and each minute divided into 60 seconds. Most angles encountered lie between 0° and 360°, although theoretically an angle can be measured from $-\infty$ to $+\infty$.

10.6 ANGLE MEASURE: RADIAN

What is the radian measure of angles?

10.6

An angle of one radian is an angle, which when placed with its vertex at the center of a circle, subtends an arc equal in length to the radius of the circle.

10.7 ANGLE MEASURE: RADIAN

Why is the radian system of measure important in mathematics?

10.7

Primarily because 1) in the calculus, certain basic relations of the trigonometric functions depend upon the use of the radian measure and 2) in mechanics, radian measure is useful in considering the motion of a particle on a circle 3) for use in certain geometric measures.

10.8 ANGLE MEASURE: CONVERSION FACTOR

What is the conversion factor used in converting from the sexagesimal to the radian system?

10.8

As one revolution is $360°$ or 2π radians, $1° = \pi/180$ radians. Therefore, multiply the angle in degrees by $\pi/180$ to convert the radians. This is also approx. equal to 0.017. (See also a set of standard tables).

10.9 ANGLE MEASURE: CONVERSION

What is the conversion factor used in converting from radian measure to sexagesimal measure?

10.9

As 2π radians = 360°, this factor is $180/\pi$ or approx. 57°.3. That is, multiply the angle in radians by $180/\pi$ to convert to degrees. (See also a set of standard tables).

10.10 ANGLE MEASURE: RADIAN

Convert the following to radian measure: 30°, 120°, 700°.

10.10

$$30^\circ = 30 \cdot \frac{\pi}{180} \text{ rad.} = \frac{\pi}{6} \text{ rad.}$$

$$120^\circ = 120 \cdot \frac{\pi}{180} \text{ rad.} = \frac{2\pi}{6} \text{ rad.}$$

$$700^\circ = 700 \cdot \frac{\pi}{180} \text{ rad.} = \frac{35\pi}{9} \text{ rad.}$$

10.11 TRIG: ANGLE MEASURE

Convert the following to sexagesimal measure: π radius, $\frac{\pi}{4}$ radius, .22 radius.

10.11

$$\pi \text{ rad.} = \pi \cdot \frac{180^\circ}{\pi} = 180^\circ$$

$$\frac{\pi}{4} \text{ rad.} = \frac{\pi}{4} \cdot \frac{180^\circ}{\pi} = 45^\circ$$

$$.22 \text{ rad.} = .22(57.3^\circ) = 12.6^\circ \text{ (approx.)}$$

What relationship exists between the length of an arc of a circle and the radius of the circle?

10.12

Let s = length of arc
r = radius
θ = central angle (radians)
then $s = r\theta$.

Note: This holds only if θ is in radians.

10.13 TRIG. DEFINITION: LENGTH OF ARC

What is the length of arc on a circle of radius 4 which is subtended by a central angle of:

a) 1 radian?

b) $135°$?

10.13

a) $S = r\Theta = 4 \cdot 1 = 4$

b) $S = r\Theta = 4 \cdot 135 \cdot \frac{\pi}{180} = 3\pi$

10.14 TRIG. DEFINITION: COORDINATE SYSTEM

What is meant by a rectangular coordinate system?

10.14

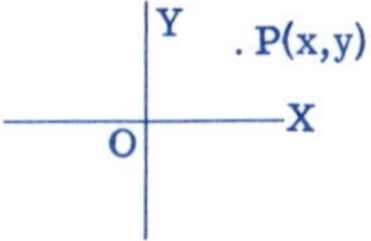

A system wherein any point in a plane is referred to two mutually perpendicular axes (X and Y axes). Using a convenient unit of measure, the point is now located by two numbers (called its coordinates) which measure its distance from each axis.

10.15 RECTANGULAR COORDINATES

What are the abscissa and ordinate of a point referred to in a rectangular coordinate system?

10.15

Abscissa is the x-coordinate or distance from the Y axis; ordinate is the y-coordinate or distance from the X axis.

10.16 RECTANGULAR COORDINATES

Locate the points A(-2, 1) and B (5, -4) on a rectangular coordinate system.

10.16

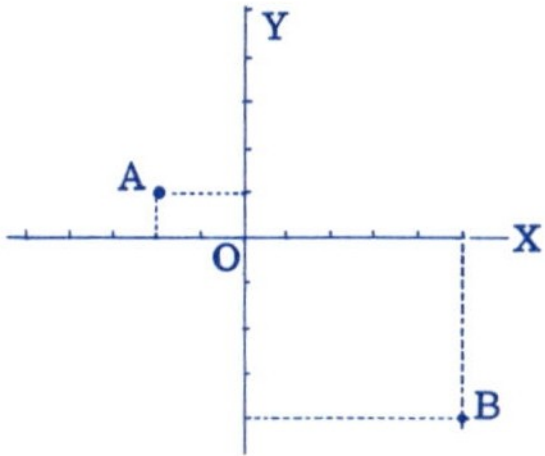

10.17 QUADRANT

What is meant by a quadrant in a rectangular coordinate system?

10.17

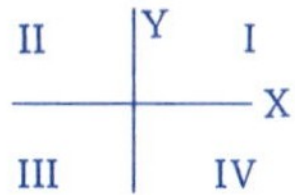

A quadrant is one of the four parts into which the plane is divided by the two axes. These are designated I, II, III or IV.

10.18 DISTANCE FORMULA

What is the distance of any point P (x, y) from the intersection of the axes O (called the origin)?

10.18 By the Pythagorean Theorem: $\overline{OP}^2 = x^2 + y^2$

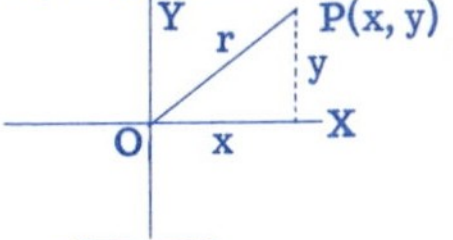

$$\overline{OP} = r = \sqrt{x^2 + y^2}$$

Note: The radical $\sqrt{x^2 + y^2}$ is always to be taken as a positive quantity.

10.19 DISTANCE FORMULA

What is the distance r from the origin to the points A (-2, 1) and B (5, -4)?

10.19

$$OA = r_1 = \sqrt{(-2)^2 + 1^2} = \sqrt{5}$$

$$OB = r_2 = \sqrt{5^2 + (-4)^2} = \sqrt{41}$$

10.20 ANGLE: STANDARD POSITION

What is meant by an angle in standard position?

10.20

An angle in standard position is one superimposed on a rectangular coordinate system so that its vertex is at the origin and its initial side lies along the positive x-axis.

11.1 TRIG. RATIO

Explain what is meant by a trigonometric ratio.

11.1

If an angle θ is placed in standard position, and a point P selected on its terminal side, the trigonometric ratios of θ are expressed as relationships between the abcissa x, ordinate y, and distance r of P.

TRIG. RATIO: SIN

Define sine θ (written $\sin \theta$).

11.2

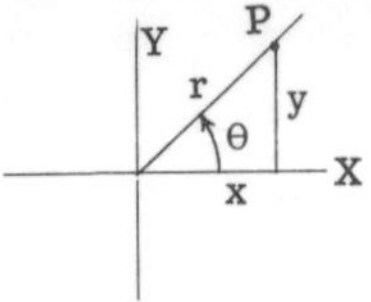

$$\sin\theta = \frac{\text{ordinate}}{\text{distance}} = \frac{y}{r}$$

11.3 TRIG. RATIOS: COS

Define cosine θ (written $\cos\theta$)

11.3

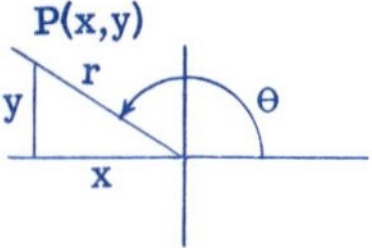

$$\cos\theta = \frac{\text{abscissa}}{\text{distance}} = \frac{x}{r}\,.$$

Define tangent θ (written $\tan \theta$).

11.4

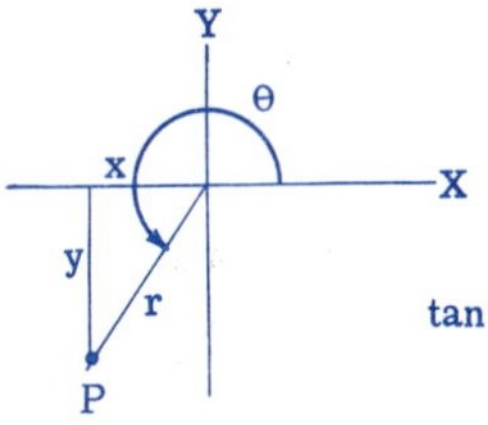

$$\tan\theta = \frac{\text{ordinate}}{\text{abscissa}} = \frac{y}{x}$$

11.5 TRIG. RATIOS: COT

Define cotangent θ (written $\cot\theta$)

11.5

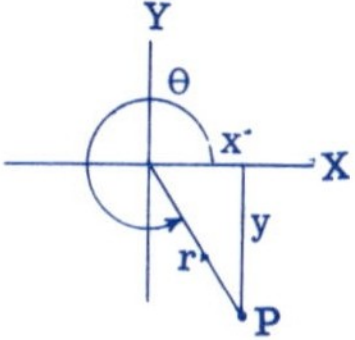

$$\cot\theta = \frac{\text{abscissa}}{\text{ordinate}} = \frac{x}{y}$$

11.6 TRIG. RATIOS: SEC

Define secant θ (written $\sec\theta$).

11.6

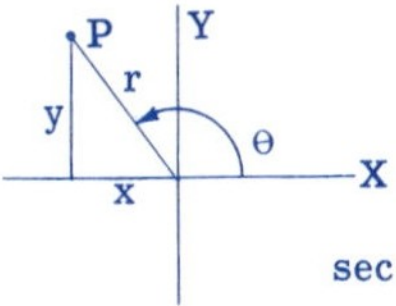

$$\sec\theta = \frac{\text{distance}}{\text{abscissa}} = \frac{r}{x}$$

11.7 TRIG. RATIO: CSC

Define cosecant θ (written csc θ).

11.7

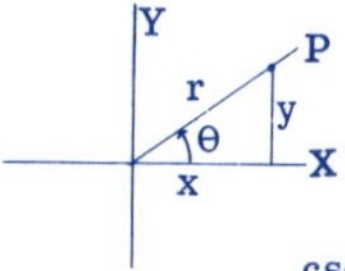

$$\csc\theta = \frac{\text{distance}}{\text{ordinate}} = \frac{r}{y}$$

11.8 TRIG. RATIOS - COTERMINAL ANGLES

What are coterminal angles?

11.8

Coterminal angles are angles which, in standard position, have the same terminal side. Consequently their trigonometric ratios will also be the same. For example, $\sin 30^\circ = \sin 390^\circ = \sin(-330^\circ)$.

What is the range of values possible for $\sin\theta$?

11.9

$$-1 \leq \sin\theta \leq +1.$$

What is the range of values possible for $\cos \theta$?

11.10

$$-1 \leq \cos\theta \leq +1$$

TRIG. RATIOS: TAN

What is the range of values possible for $\tan\theta$?

11.11

$$-\infty < \tan\theta < +\infty$$

TRIG. RATIOS: COT

What is the range of values possible for $\cot \theta$?

11.12

$$-\infty < \cot\theta < +\infty$$

11.13 TRIG. RATIO: SEC

What is the range of values possible for $\sec\theta$?

11.13

$$\sec\theta \geq 1 \text{ or } \sec\theta \leq -1$$

11.14 TRIG. RATIO: COS

What is the range of values possible for $\cos\theta$?

11.14

$$\csc\theta \geq 1 \text{ or } \csc\theta \leq -1$$

11.15 TRIG. RATIOS: SIGNS

Make a table showing the signs of the trig. ratios in the four quadrants.

11.15

	Q I	Q II	Q III	Q IV
sin θ csc θ	+	+	-	-
cos θ sec θ	+	-	-	+
tan θ cot θ	+	-	+	-

Name the quadrants in which the sine and cosecant have opposite signs.

11.16

None, as the sine and cosecant always have the same sign.

Name the quadrants in which the sine and tangent have the same sign.

11.17

Quadrants I and IV.

Name the quadrant in which the sine and cosine are both negative.

11.18

Quadrant III.

Name the quadrant in which the cosine and tangent are both negative.

11.19

Quadrant II.

Name the quadrant in which the cosine is positive and the tangent negative.

11.20

Quadrant IV.

Name the quadrant in which the sine is positive and the tangent negative.

11.21

Quadrant II.

Find the trigonometric ratios of an angle θ in standard position having its terminal side passing through the point (12, -5).

11.22

$\sin\theta = \frac{-5}{13}$

$\cos\theta = \frac{12}{13}$

$\tan\theta = \frac{-5}{12}$

$\cot\theta = \frac{12}{-5}$

$\sec\theta = \frac{13}{12}$

$\csc\theta = \frac{13}{-5}$

$x = 12$

$y = -5$

$r = 13$

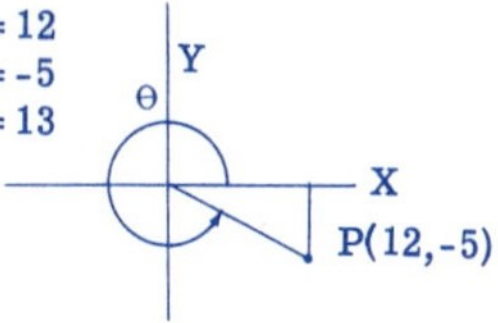

11.23 TRIG. RATIOS: RECIPROCAL

What are the three pairs of reciprocal trigonometric ratios?

11.23

$$\sin\theta \csc\theta = 1.$$
$$\cos\theta \sec\theta = 1.$$
$$\tan\theta \cot\theta = 1.$$

What are the values of $\tan\theta$ and $\cot\theta$ in terms of $\sin\theta$ and $\cos\theta$?

11.24

$$\tan\Theta = \frac{y}{x} = \frac{\frac{y}{r}}{\frac{x}{r}} = \frac{\sin\Theta}{\cos\Theta} \qquad \cot\Theta = \frac{x}{y} = \frac{\frac{x}{r}}{\frac{y}{r}} = \frac{\cos\Theta}{\sin\Theta}$$

Given $\sin\theta = a$, find $\cos\theta$ and $\tan\theta$.

11.25

1) Construct an angle in standard position showing $\sin\theta = a$.

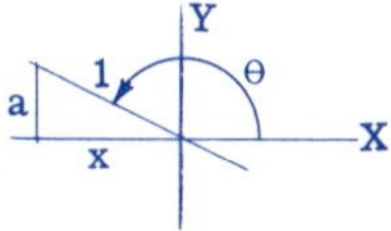

2) $x = \pm\sqrt{1 - a^2}$

3) $\cos\theta = \pm\sqrt{1 - a^2}$ $\quad \tan\theta = \pm\dfrac{a}{\sqrt{1 - a^2}}$

11.26

TRIG. RATIOS

Given $\sin\theta = \frac{5}{13}$ and $\cos\theta = -\frac{12}{13}$,

find the value of $\frac{\tan\theta}{\sec\theta + \csc\theta}$

11.26

θ is in Quadrant II

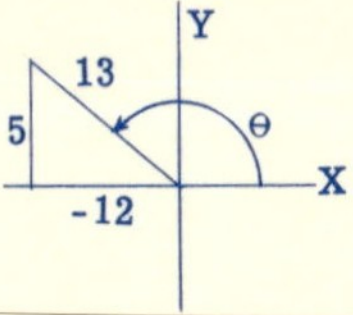

$$\frac{\tan\theta}{\sec\theta+\csc\theta}=\frac{-5/12}{-13/12+13/5}=\frac{-5/12}{13(7/60)}$$

$$=-\frac{5/12}{91/60}=-\frac{5}{12}\cdot\frac{60}{91}=-\frac{25}{91}$$

TRIG. RATIOS

Given $\tan\theta = \frac{1}{3}$ (θ in Quadrant III) find the value of

$$\frac{\tan\theta - \sec\theta}{\cos\theta + \sin\theta}$$

11.27

$$\frac{\tan\theta - \sec\theta}{\cos\theta + \sin\theta} = \frac{\frac{1}{3} - \frac{\sqrt{10}}{-3}}{\frac{-3}{\sqrt{10}} + \frac{-1}{\sqrt{10}}} = \frac{\frac{1+\sqrt{10}}{3}}{-\frac{4}{\sqrt{10}}}$$

$$= \frac{1+\sqrt{10}}{3} \cdot -\left(\frac{\sqrt{10}}{4}\right) = -\frac{10+\sqrt{10}}{12}.$$

$r = \sqrt{10}$

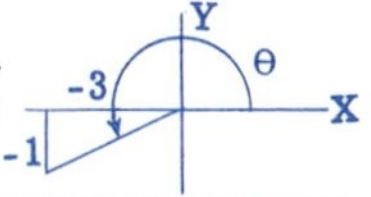

Find the trigonometric ratios for $\theta = 30°$.

11.28

From geometry, a 30°- 60° right Δ has sides in ratio 1, $\sqrt{3}$, 2. Therefore, for $\theta = 30°$; $x = \sqrt{3}$, $y = 1$, $r = 2$.

$\sin 30° = \frac{1}{2}$, $\cos 30° = \frac{\sqrt{3}}{2}$, $\tan 30° = \frac{1}{\sqrt{3}}$

$\cot 30° = \frac{\sqrt{3}}{1}$, $\sec 30° = \frac{2}{\sqrt{3}}$, $\csc 30° = 2$

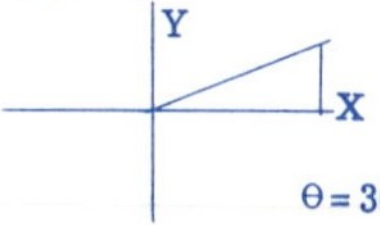

$\theta = 30°$

List the angles $0^\circ \leq \theta \leq 360^\circ$ for which the sides x, y and r are in numerical ratio to

$$1, \sqrt{3} \text{ and } 2; \text{ that is } \frac{x}{1} = \frac{y}{\sqrt{3}} = \frac{r}{2} \text{ or } \frac{x}{\sqrt{3}} = \frac{y}{1} = \frac{r}{2}$$

11.29

$$\theta = 30^\circ, 60^\circ, 120^\circ, 150^\circ, 210^\circ, 240^\circ, 300^\circ, 330^\circ.$$

Find the trigonometric ratios for $\theta = 330^\circ$.

11.30

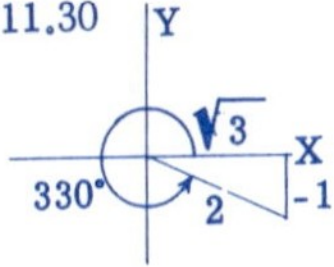

$\sin 330° = -\frac{1}{2}$

$\cos 330° = +\frac{\sqrt{3}}{2}$

$\tan 330° = -\frac{1}{\sqrt{3}}$

$\cot 330° = -\sqrt{3}$

$\sec 330° = +2/\sqrt{3}$

$\csc 330° = -2$

11.31 TRIG. RATIOS: 45

Find the trigonometric ratio for $\theta = 45^\circ$.

11.31

This triangle is an isosceles right triangle, and the sides are in ratio, 1, 1, $\sqrt{2}$.

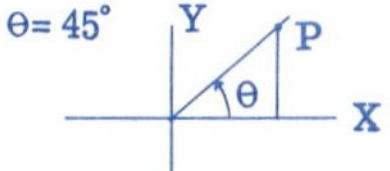

$\sin\theta = \cos\theta = \frac{1}{\sqrt{2}}$; $\tan\theta = \cot\theta = 1$; $\sec\theta = \csc\theta = \sqrt{2}$.

List the angles $0 \leq \theta \leq 360°$ for which the sides x, y and r numerically have the ratios 1, 1, $\sqrt{2}$.

11.32

$$\theta = 45^\circ, 135^\circ, 225^\circ, 315^\circ.$$

11.33 TRIG. RATIOS: QUADRANTAL ANGLE

What is meant by a quadrantal angle?

11.33

An angle such as 0°, 90°, 180°, or 270° whose terminal side falls on one of the coordinate axes.

TRIG. RATIOS: 0°

Find the trigonometric ratios for $\theta = 0^\circ$

11.34 We let θ be very small. Then as $\theta \to 0^\circ$, $y \to 0$, $x \to r$.

Then $\sin 0^\circ = \frac{y}{r} = 0$ $\quad \csc 0^\circ = \frac{r}{y} = \infty$

$\cos 0^\circ = \frac{x}{r} = 1$ $\quad \sec 0^\circ = \frac{r}{x} = 1$

$\tan 0^\circ = \frac{y}{x} = 0$ $\quad \cot 0^\circ = \frac{x}{y} = \infty$

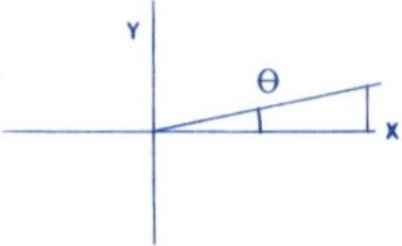

Note: By ∞ we mean as $\theta \to 0$, the ratio gets larger and larger, without limit.

11.35 TRIG. RATIOS: QUADRANTAL ANGLES

Make a table showing the trigonometric ratios for 0°, 90°, 180°, 270°.

11.35

Θ	$\sin\Theta$	$\cos\Theta$	$\tan\Theta$	$\cot\Theta$	$\sec\Theta$	$\csc\Theta$
0°	0	1	0	∞	1	∞
90°	1	0	∞	0	∞	1
180°	0	-1	0	∞	-1	∞
270°	-1	0	∞	0	∞	-1

Find the value of $\dfrac{\cos 0^\circ - \sin 30^\circ - \tan 45^\circ}{\cot^2 60^\circ - \sin 90^\circ + \sec^2 30^\circ}$

11.36

$$\frac{1-\frac{1}{2}-1}{\frac{1}{3}-1+\frac{4}{3}}=\frac{-\frac{1}{2}}{\frac{2}{3}}=-\frac{3}{4}$$

Find the value of $\dfrac{4\tan\frac{5\pi}{6} - 2\sec\frac{11\pi}{6}}{3\sin\frac{3\pi}{2} - 5\cos\pi}$

11.37

$$\frac{4\tan 150^\circ - 2\sec 330^\circ}{3\sin 270^\circ - 5\cos 180^\circ} = \frac{4\left(-\frac{1}{\sqrt{3}}\right) - 2\left(\frac{2}{\sqrt{3}}\right)}{3(-1) - 5(-1)} = \frac{-\frac{8}{\sqrt{3}}}{2} = -\frac{4}{\sqrt{3}}$$

11.38 TRIG. RATIOS

Solve for x: $x \tan\frac{2\pi}{3} = \sin\frac{\pi}{4}\cos\frac{5\pi}{4} + \cos\frac{\pi}{4}\sin\frac{5\pi}{4}$

11.38 $x \cdot \tan 120^\circ = \sin 45^\circ \cdot \cos 225^\circ + \cos 45^\circ \cdot \sin 225^\circ$

$$x(-\sqrt{3}) = \frac{1}{\sqrt{2}} \cdot \frac{-1}{\sqrt{2}} + \frac{1}{\sqrt{2}} \cdot \frac{-1}{\sqrt{2}}$$

$$x(-\sqrt{3}) = -1$$

$$x = \frac{-1}{-\sqrt{3}} = \frac{1}{\sqrt{3}}$$

12.1 RIGHT TRIANGLE SOLUTION

What special conditions govern the trigonometric ratios of an acute angle in a right triangle?

12.1

As a right triangle will have only acute angles, the ratios will always be positive. Therefore one can replace y, x and r by the "side opposite the angle", the "side adjacent to the angle" and the "hypotenuse".

12.2 RIGHT TRIANGLE SOLUTION

Define the trigonometric ratios for an acute angle in a right angle.

12.2

$\sin A = a/c$	$\csc A = c/a$
$\cos A = b/c$	$\sec A = c/b$
$\tan A = a/b$	$\cot A = b/a$

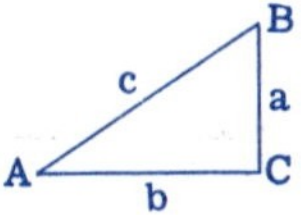

Note: These hold only for an acute angle in a right triangle.

12.3 RIGHT TRIANGLE SOLUTION

What is a complementary angle?

12.3

Two angles are complementary if their sum is 90°. Therefore, the two acute angles in a right triangle are complementary angles.

12.4 RIGHT TRIANGLE SOLUTION

What relations hold for the trigonometric ratios of complementary angles?

12.4

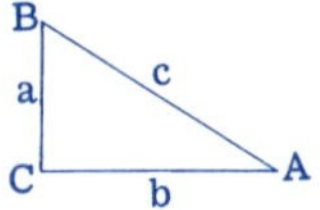

$\sin A = \cos B$; $\cos A = \sin B$; $\tan A = \cot B$; $\cot A = \tan B$; etc. That is, any function of an acute angle is equal to the cofunction of its complement.

Express the following as functions of their complementary angles: $\cos 65^\circ$, $\tan \frac{\pi}{3}$, $\sin 3\theta$, $\cot 25^\circ\ 15'$.

12.5

$$\cos 65 = \sin 25°$$

$$\tan\frac{\pi}{3} = \cot\ \frac{\pi}{2} - \frac{\pi}{3}\ = \cot\frac{\pi}{6}$$

$$\sin 3\theta = \cos(90° - 3\theta)$$

$$\cot 25°\,15' = \tan 64°\,45'$$

12.6 RIGHT TRIANGLE SOLUTION

Find θ if $\tan 75^\circ = \cot \theta$, θ acute.

12.6

a) $\tan 75^\circ = \cot \theta$

$\cot 15^\circ = \cot \theta$

$\therefore \theta = 15^\circ$

b) If $\tan 75^\circ = \cot \theta$

then $75^\circ + \theta = 90^\circ$

$\therefore \theta = 15^\circ$

12.7 RIGHT TRIANGLE SOLUTION

Find θ if $\sin\theta = \cos 2\theta$, θ acute.

12.7

a) $\sin\theta = \cos 20^\circ = \sin 70^\circ \therefore \theta = 70^\circ$

b) or if $\sin\theta = \cos 20^\circ$ then $\theta + 20^\circ = 90^\circ \therefore \theta = 70^\circ$

12.8 SOLUTION OF RIGHT TRIANGLES

If θ is an acute angle less than $60°$, show that $\sin(60° - \theta) = \cos(30° + \theta)$.

12.8

$$\sin(60^\circ - \Theta) = \cos\left[90^\circ - (60^\circ - \Theta)\right] = \cos(30^\circ + \Theta)$$

$$\therefore \sin(60^\circ - \Theta) = \cos(30^\circ + \Theta)$$

12.9 RIGHT TRIANGLE SOLUTION

What must be known to solve a right triangle?

12.9

A right triangle has 6 parts; three sides, two acute angles and one right angle. In order to solve for missing parts, one must know

(1) two sides, or

(2) one acute angle and one side.

12.10 RIGHT TRIANGLE SOLUTION

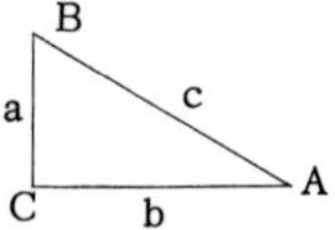

Outline the solution for missing parts in the right triangle ABC, if A and b are known.

12.10

$$\tan A = \frac{a}{b} \text{ or } a = b \tan A$$

$$\cos A = \frac{b}{c} \text{ or } c = \frac{b}{\cos A}$$

$$B = 90^\circ - A$$

12.11 RIGHT TRIANGLE SOLUTION

Outline the solution of missing parts in the right triangle ABC, if a and b are known.

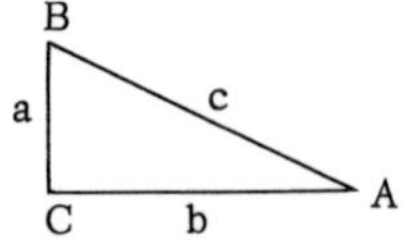

12.11

$$\tan A = \frac{a}{b} \qquad \tan B = \frac{b}{a}$$

$$c = \sqrt{a^2 + b^2} \text{ or } c = \frac{a}{\sin A} = \frac{b}{\sin B}$$

12.12 RIGHT TRIANGLE SOLUTION

If $\tan B = \frac{3}{2}$ and $c = 2\sqrt{13}$, find a and b.

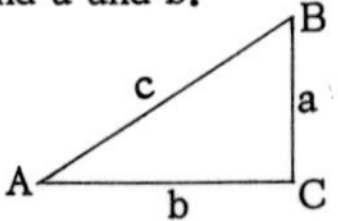

12.12

$$\tan B = \frac{3}{2}, \text{ therefore } \sin B = \frac{3}{\sqrt{13}} \text{ and } \cos B = \frac{2}{\sqrt{13}}$$

$$a = c \cos B = 2\sqrt{13} \cdot \frac{2}{\sqrt{13}} = 4$$

$$b = c \sin B = 2\sqrt{13} \cdot \frac{3}{\sqrt{13}} = 6$$

12.13 RIGHT TRIANGLE SOLUTION

Find a and b in terms of θ and m.

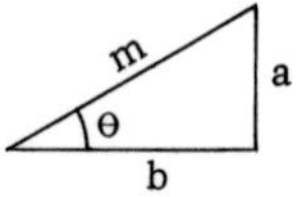

12.13

$$a = m \sin\theta$$
$$b = m \cos\theta$$

12.14 RIGHT TRIANGLE SOLUTION

Find sin A, cos A and tan A in the right triangle ABC.

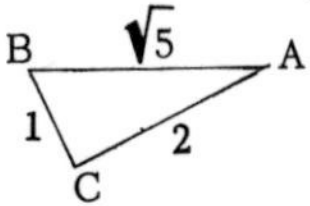

12.14

$$\sin A = \frac{1}{\sqrt{5}}$$

$$\cos A = \frac{2}{\sqrt{5}}$$

$$\tan A = \frac{1}{2}$$

Give the length of CD in terms of p, α and β.

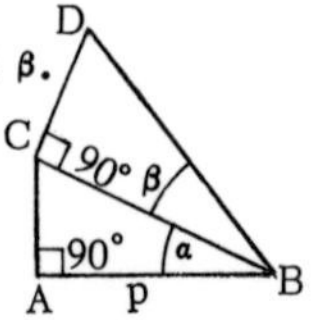

12.15

$$\begin{aligned} &\text{In } \Delta ABC, && (CB) = p \sec \alpha \\ &\text{In } \Delta BCD, && (CD) = (CB) \tan \beta \\ &\therefore && (CD) = p \sec \alpha \tan \beta \end{aligned}$$

12.16 RIGHT TRIANGLE SOLUTION

Find the angle that the diagonal of a cube makes with the diagonal of one of the faces of the cube.

12.16 Let edge of cube be a. Find angle AOB ($\triangle$AOB is a right triangle with $\angle$OAB = 90°).

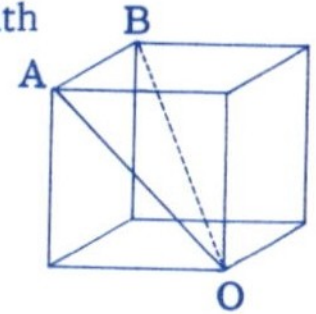

$$OA = \sqrt{a^2 + a^2} = a\sqrt{2}$$

$$\therefore \tan\ AOB = \frac{AB}{OA} = \frac{a}{a\sqrt{2}} = \frac{1}{\sqrt{2}}$$

$\angle AOB = 35.3°$ (by tables or slide rule)

12.17 RIGHT TRIANGLE SOLUTION

Solve the right triangle ABC, if a=48 and b=19.

12.17

$$\tan A = \frac{48}{19}; \qquad A = 68.^\circ 4$$

$$B = 90 - A = 21.^\circ 6$$

$$c = \frac{a}{\sin A} = \frac{b}{\sin B} = 51.6$$

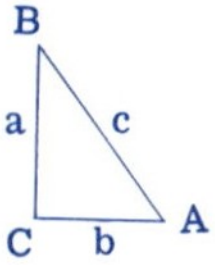

(Arithmetic done by tables or slide rule)

12.18 ANGLES: ELEVATION AND DEPRESSION

What are the angles of elevation and depression for a line of sight?

12.18

Through a point of observation O draw the line of sight O P of an object P, and a horizontal line O H; then the acute angle between these lines is an angle of elevation or depression according as P is above or below the horizontal line.

12.19 RIGHT TRIANGLE SOLUTION: PROJECTION

What is the projection of a point A upon a line L?

12.19

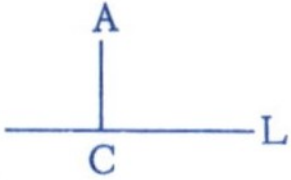

The projection of point A on the line L is the point C which is at the foot of the perpendicular from A to the line.

12.20 SOLUTION OF RIGHT TRIANGLES: PROJECTION

What is the projection of a line segment AB upon another line L?

12.20

The projection of AB on L is the line segment CD which joins the projections of the points A and B upon the line L. If AB makes an angle θ with L, the length of the projection is (length of AB) ($\cos\theta$).

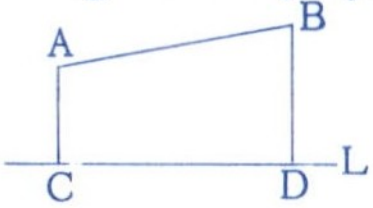

12.21 RIGHT TRIANGLE SOLUTION

If the angle of elevation of the sun is 42° 30′, how long a shadow will a man 6 ft. 2 in. tall cast on level ground?

12.21

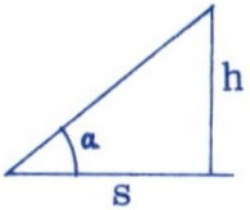

$\alpha = 42^\circ\ 30'$
$h = 6$ ft. 2 in.
$= 74$ in.
Find s

$\cot \alpha = s/h$

$\therefore s = (74 \cot 42^\circ\ 30')$ inches
$= 80.76$ inches
$= 6$ ft. 9 in. (to nearest inch)

12.22 RIGHT TRIANGLE SOLUTION

From the top of a mountain, the angles of depression of two successive milestones in the horizontal plane below are α and β. Find the expression for h, the height of the mountain, in terms of α and β.

12.22

(1) $\cot(90° - \alpha) = \tan\alpha = \frac{h}{p}$; (2) $\cot(90° - \beta) = \tan\beta = \frac{h}{1+p}$

From (1) $p = \frac{h}{\tan\alpha}$; (2) $\left(1 + \frac{h}{\tan\alpha}\right)\tan\beta = h.$

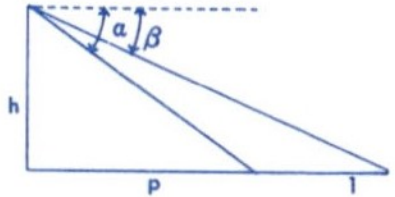

Solving for h: $h = \frac{\tan\alpha \tan\beta}{\tan\alpha - \tan\beta}$

12.23 RIGHT TRIANGLE SOLUTION

A line segment is 27 units long and makes an angle of 30° 15′ with the horizontal. Find the length of the projection of this segment upon a vertical line.

12.23

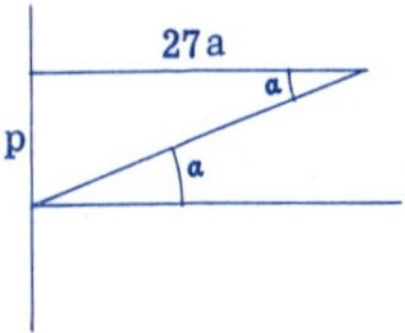

$\alpha = 30^\circ\ 15'$

$\frac{p}{27} = \sin\alpha$

$\therefore p = 27 \sin 30^\circ\ 15' = 13.60$ units

12.24 VECTOR

What is a vector?

12.24

A vector is a directed line segment whose length represents, to some convenient scale, the magnitude of a vector quantity. A vector quantity is a quantity which has both magnitude and direction, such as a force, velocity or acceleration.

12.25 RIGHT TRIANGLE SOLUTION: VECTOR

What are the X and Y (or horizontal and vertical) components of a vector, and how are they obtained?

12.25

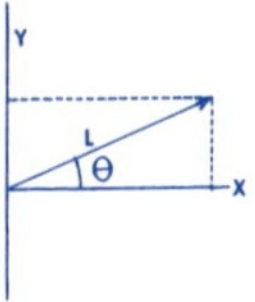

If a vector is referred to a set of X and Y axes, the components are the lengths of the projections upon those axes. If the vector has length L and makes an angle θ with the horizontal, the components are $L \cos \theta$ and $L \sin \theta$.

12.26 RIGHT TRIANGLE SOLUTION

An automobile weighing 2500 pounds is standing on a road that is inclined 10° to the horizontal. Find the components of the force parallel to the road and perpendicular to the road.

12.26

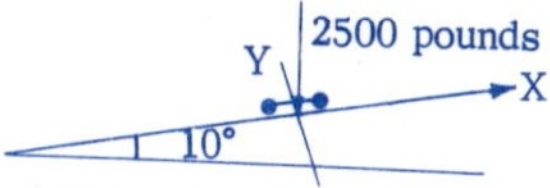

The force makes an angle of 80° with the X axis: therefore

$F_x = 2500 \cos 80° = 434$ pounds

$F_y = 2500 \sin 80° = 2462$ pounds

12.27 RIGHT TRIANGLE SOLUTION

Two forces of 125 lb. and 180 lb. act at right angles to each other. Find the magnitude and direction of their combined effect (called the resultant force).

12.27

Find R and Θ

$\tan\Theta = \frac{125}{180} = \frac{25}{36}$

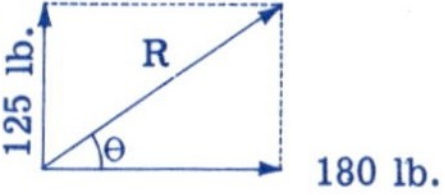

$\Theta = 34^\circ 50'$

$R = \frac{125}{\sin\Theta} = 219 \text{ lb.}$

13.1 REDUCTION FORMULA

What is meant by a reduction formula?

13.1

A reduction formula is a process whereby a function of any angle is reduced or set equal to the corresponding function of an acute angle.

What is meant by a reference angle for any angle θ placed in standard position?

13.2

A reference angle Φ is defined as the positive acute angle formed by the terminal side of θ and the x axis.

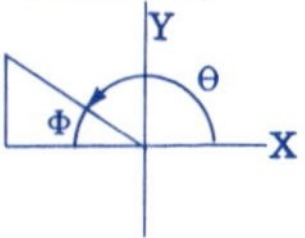

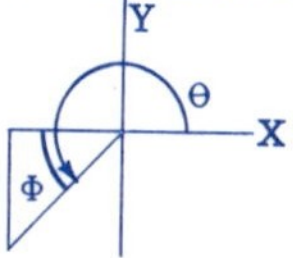

REDUCTION FORMULAS

What is a reduction formula for the trigonometric functions?

13.3

Trig. function [θ] = $\pm$ same trig. function [reference angle]; the choice of + or - depends on the sign of the original function of θ, as determined by the quadrant in which the terminal side of θ would lie.

13.4 REDUCTION FORMULAS

What are the steps in reducing a trigonometric function of any angle to a trigonometric function of an acute angle?

13.4.

(1) Find the reference angle.
(2) Use the same function as originally desired, on the reference angle.
(3) Apply the + or - sign to the function depending on whether the function of the original angle θ is + or -.

13.5 REDUCTION FORMULAS

Express the following as functions of acute angles.

a) $\tan 200°$
b) $\sin 310°$
c) $\cos 105°$

13.5

a) $\tan 200^\circ = +\tan 20^\circ$
b) $\sin 310^\circ = -\sin 50^\circ$
c) $\cos 105^\circ = -\cos 75^\circ$.

13.6 REDUCTION FORMULAS

Find the value of the following:

a) $\sin \frac{7\pi}{6}$

b) $\tan 480°$

c) $\cos 300°$

13.6

a) $\sin\frac{7\pi}{6} = -\sin\frac{\pi}{6} = -\frac{1}{2}$

b) $\tan 480° = -\tan 60° = -\sqrt{3}$

c) $\cos 300° = +\cos 60° = +\frac{1}{2}$

13.7 REDUCTION FORMULAS

What are the reduction formulas for functions of negative angles?

13.7

$$\sin(-\theta) = -\sin\theta$$

$$\cos(-\theta) = +\cos\theta$$

$$\tan(-\theta) = -\tan\theta$$

Reduce the following:

a) $\sin(-210°)$

b) $\cos(-135°)$

13.8

a) $\sin(-210^\circ) = -\sin 210^\circ = -(-\sin 30^\circ) = \sin 30^\circ = \frac{1}{2}$

b) $\cos(-135^\circ) = +\cos 135^\circ = -\cos 45^\circ = -\frac{1}{\sqrt{2}}$

13.9 REDUCTION FORMULAS

Evaluate $$\frac{\sin\frac{5\pi}{4}\tan\frac{5\pi}{3}\sec\frac{5\pi}{4}}{\cot\frac{13\pi}{6}+\csc\left(-\frac{\pi}{3}\right)}$$

13.9

$$\frac{\left(-\sin\frac{\pi}{4}\right)\left(-\tan\frac{\pi}{3}\right)\left(-\sec\frac{\pi}{4}\right)}{\cot\frac{\pi}{6} - \csc\frac{\pi}{3}} = \frac{\frac{1}{\sqrt{2}}(-\sqrt{3})(-\sqrt{2})}{\sqrt{3} - \frac{2}{\sqrt{3}}} = \frac{-\sqrt{3}}{\frac{1}{\sqrt{3}}} = -3$$

13.10 REDUCTION FORMULAS

Given $\sin 65^\circ = a$, find the value of $\sin 245^\circ$ and $\tan 205^\circ$.

13.10

$$\sin 245^\circ = -\sin 65^\circ = -a$$

$$\tan 205^\circ = +\tan 25^\circ = \cot 65^\circ = \frac{\sqrt{1-a^2}}{a}$$

13.11 REDUCTION FORMULAS

Simplify $\dfrac{\sin(180° - \theta)}{\sin(180° + \theta)}$

13.11

$$\frac{\sin(180^\circ - \theta)}{\sin(180^\circ + \theta)} \equiv \frac{\sin\theta}{-\sin\theta} \equiv -1$$

13.12 REDUCTION FORMULAS

Simplify $\sin(-\theta) \cdot \cot(180^\circ - \theta)$.

13.12

$$\sin(-\theta)\cot(180^\circ - \theta) \equiv (-\sin\theta)(-\cot\theta) \equiv +\cos\theta,$$

provided $\theta \neq n\pi$ $(n = 0, 1, 2, \ldots)$

13.13 REDUCTION FORMULAS

Simplify $\dfrac{\sin(-\theta)\cos(-\theta)}{\sec(\theta-\pi)\csc(\theta+\pi)}$

13.13

$$\frac{\sin(-\theta)\cos(-\theta)}{\sec(\theta-\pi)\csc(\theta+\pi)} \equiv \frac{(-\sin\theta)(\cos\theta)}{(-\sec\theta)(-\csc\theta)} \equiv -\frac{\sin\theta\cos\theta}{\sec\theta\csc\theta}$$

$$\equiv -\sin^2\theta\cos^2\theta .$$

14.1 TRIG. IDENTITY

What is a trigonometric identity?

14.1

An equality dealing with trigonometric functions which is true for all values of the angles involved. (Let $\equiv$ represent an identity hereafter).

14.2

TRIG. IDENTITY

There are eight fundamental trig. identities. Name the three reciprocal relations.

14.2

$$\sin\theta\csc\theta \equiv 1 \text{ or } \csc\theta \equiv \frac{1}{\sin\theta}$$

$$\cos\theta\sec\theta \equiv 1 \text{ or } \sec\theta \equiv \frac{1}{\cos\theta}$$

$$\cot\theta\tan\theta \equiv 1 \text{ or} \cot\theta \equiv \frac{1}{\tan\theta}$$

14.3 TRIG. IDENTITY

Two of the fundamental identities express the tangent and cotangent in terms of the sine and cosine. Name them.

14.3

$$\tan\Theta \equiv \frac{\sin\Theta}{\cos\Theta}$$

$$\cot\Theta \equiv \frac{\cos\Theta}{\sin\Theta}$$

14.4 IDENTITY: SQUARES OF FUNCTIONS

Three of the fundamental identities involve the squares of the functions. Name them.

14.4

$$\sin^2\Theta + \cos^2\Theta \equiv 1$$

$$1 + \tan^2\Theta \equiv \sec^2\Theta$$

$$1 + \cot^2 \quad \equiv \csc^2\Theta$$

IDENTITIES - USE OF

What are some general principles for simplifying trigonometric expressions?

14.5

(1) Use algebraic simplification.
(2) Use fundamental identities.
(3) If all else fails, express all functions in terms of sines and cosines.
(4) Try to avoid possible ambiguity of sign resulting from the use of radicals.

14.6 IDENTITIES: USE OF

Simplify $\sin\theta \cos\theta \tan\theta$.

14.6

$$\sin\theta\cos\theta\tan\theta \equiv \sin\theta\cos\theta\frac{\sin\theta}{\cos\theta} \equiv \sin^2\theta\ .$$

14.7 IDENTITIES: USE OF

Simplify $\dfrac{1 - \sin^2\theta}{1 - \cos^2\theta}$

14.7

$$\frac{1 - \sin^2\Theta}{1 - \cos^2\Theta} \equiv \frac{\cos^2\Theta}{\sin^2\Theta} \equiv \cot^2\Theta$$

o

14.8 IDENTITIES - USE OF

Simplify $\dfrac{\cot\theta - \cos\theta}{1 - \csc\theta}$

14.8

$$\frac{\cot\theta - \cos\theta}{1 - \csc\theta} \equiv \frac{\dfrac{\cos\theta}{\sin\theta} - \cos\theta}{1 - \dfrac{1}{\sin\theta}} \equiv \frac{\dfrac{\cos\theta(1-\sin\theta)}{\sin\theta}}{\dfrac{\sin\theta - 1}{\sin\theta}} \equiv$$

$$\frac{\cos\theta(1-\sin\theta)}{\sin\theta - 1} \equiv -\cos\theta$$

14.9 TRIG. IDENTITIES: USE OF

Simplify: $\frac{1}{1+\cos\theta}+\frac{1}{1-\cos\theta}$

14.9

$$\frac{1}{1+\cos\theta}+\frac{1}{1-\cos\theta}\equiv\frac{(1-\cos\theta)+(1+\cos\theta)}{1-\cos^2\theta}\equiv\frac{2}{\sin^2\theta}\equiv 2\csc^2\theta$$

14.10 IDENTITIES: USE OF

Simplify $\frac{\sin\theta}{\csc\theta - \cot\theta} + \frac{\sin\theta}{\csc\theta + \cot\theta}$

14.10

$$\frac{\sin\theta}{\frac{1}{\sin\theta}-\frac{\cos\theta}{\sin\theta}}+\frac{\sin\theta}{\frac{1}{\sin\theta}+\frac{\cos\theta}{\sin\theta}}\equiv\frac{\sin^2\theta}{1-\cos\theta}+\frac{\sin^2\theta}{1+\cos\theta}$$

$$\equiv\sin^2\theta\left[\frac{(1+\cos\theta)+(1-\cos\theta)}{1-\cos^2\theta}\right]\equiv\sin^2\theta\left[\frac{2}{\sin^2\theta}\right]\equiv 2$$

14.11 FUNDAMENTAL IDENTITIES

$$\text{Simplify } \frac{\sin^3\theta - \cos^3\theta}{\sin\theta - \cos\theta}$$

14.11

$$\frac{\sin^3\theta - \cos^3\theta}{\sin\theta - \cos\theta} \equiv \frac{\cancel{(\sin\theta - \cos\theta)}(\sin^2\theta + \sin\theta\cos\theta + \cos^2\theta)}{\cancel{(\sin\theta - \cos\theta)}}$$

$$\equiv 1 + \sin\theta\cos\theta.$$

14.12 IDENTITIES: USE OF

$$\text{Simplify } \frac{\sqrt{1+\tan^2\theta}}{\sec\theta} + \frac{\sqrt{1+\cot^2\theta}}{\csc\theta}$$

14.12

$$\frac{\sqrt{1+\tan^2\theta}}{\sec\theta}+\frac{\sqrt{1+\cot^2\theta}}{\csc\theta}\equiv\frac{|\sec\theta|}{\sec\theta}+\frac{|\csc\theta|}{\csc\theta}=\begin{array}{l}1+1=2\ (\theta \text{ in QI})\\ -1+1=0\ (\theta \text{ in QII})\\ -1-1=-2\ (\theta \text{ in QIII})\\ +1-1=0\ (\theta \text{ in QIV})\end{array}$$

14.13 TRIG. IDENTITIES: USE OF

Simplify $\sec^2 \theta \sec^2 \Phi - \tan^2 \theta \sec^2 \Phi - \tan^2 \Phi$

14.13

$$\sec^2\theta\sec^2\phi - \tan^2\theta\sec^2\phi - \tan^2\phi \equiv \sec^2\phi(\sec^2\theta - \tan^2\theta)$$
$$- \tan^2\phi \equiv \sec^2\phi\cdot 1 - \tan^2\phi \equiv 1$$

14.14 TRIG. IDENTITY: USE OF

Simplify $\dfrac{\sin^2 \theta}{\cos\theta(1-\sec\theta)(1+\cos\theta)}$

14.14

$$\frac{1-\cos^2\Theta}{\cos\Theta(1-\sec\Theta)(1+\cos\Theta)} \equiv \frac{1-\cos\Theta}{\cos\Theta(1-\sec\Theta)} \equiv \frac{1-\cos\Theta}{\cos\Theta-1} \equiv -1$$

Note: Only true when $\Theta \neq 0$.

TRIG. IDENTITY

How should identities be proven?

14.15

(1) Convert left member to right member.
(2) Convert right member to left member.
(3) Convert both members independently to the same result.

Note: DO NOT treat the identity as an equation, and "transpose" members from one side to the other.

14.16 TRIG. IDENTITY

Prove the identity $\tan\theta + \cot\theta = \sec\theta \csc\theta$

14.16

$$\tan\theta + \cot\theta \equiv \frac{\sin\theta}{\cos\theta} + \frac{\cos\theta}{\sin\theta} \equiv \frac{\sin^2\theta + \cos^2\theta}{\sin\theta\cos\theta}$$

$$\equiv \frac{1}{\sin\theta\cos\theta} \equiv \sec\theta\csc\theta$$

14.17 TRIG. IDENTITY

Prove the identity $\tan\theta + \dfrac{\cos\theta}{1+\sin\theta} \equiv \sec\theta$

14.17

$$\tan\theta + \frac{\cos\theta}{1+\sin\theta} \equiv \frac{\sin\theta}{\cos\theta} + \frac{\cos\theta}{1+\sin\theta} \equiv \frac{\sin\theta(1+\sin\theta)+\cos^2\theta}{\cos\theta(1+\sin\theta)}$$

$$\equiv \frac{\sin\theta + 1}{\cos\theta(1+\sin\theta)} \equiv \frac{1}{\cos\theta} \equiv \sec\theta$$

TRIG. IDENTITY

Prove the identity $\cos\theta\,(1-\tan\theta)\,(\sin\theta+\cos\theta) \equiv 2\cos^2\theta - 1$

14.18

$$
\begin{aligned}
\cos\theta(1-\tan\theta)(\sin\theta+\cos\theta) &\equiv \left[\cos\theta-\sin\theta\right]\left[\sin\theta+\cos\theta\right] \\
&\equiv \cos^2\theta-\sin^2\theta \\
&\equiv \cos^2\theta-(1-\cos^2\theta) \\
&\equiv 2\cos^2\theta-1
\end{aligned}
$$

14.19 TRIG. IDENTITY

Prove the identity $\sec\theta\cot\theta + \dfrac{1}{\sin\theta} = 2\csc\theta$

14.19

$$\sec\theta\cot\theta + \frac{1}{\sin\theta} \equiv \frac{1}{\cancel{\cos\theta}}\cdot\frac{\cancel{\cos\theta}}{\sin\theta} + \frac{1}{\sin\theta} \equiv \frac{2}{\sin\theta} \equiv 2\csc\theta .$$

14.20 TRIG. IDENTITY

Prove the identity $\sec^2\theta - \cos^2\theta \equiv \tan^2\theta + \sin^2\theta$

14.20

$$\sec^2\theta - \cos^2\theta \equiv \frac{1}{\cos^2\theta} - \cos^2\theta \equiv \frac{1-\cos^4\theta}{\cos^2\theta} \equiv$$

$$\frac{(1+\cos^2\theta)(1-\cos^2\theta)}{\cos^2\theta} \equiv \left(\frac{1}{\cos^2\theta}+1\right)\left(\sin^2\theta\right) \equiv \tan^2\theta + \sin^2\theta$$

14.21 TRIG. IDENTITY

Prove the identity:

$$\frac{\tan\theta}{1+\csc\theta}+\frac{\tan\theta}{1+\sin\theta}\equiv\tan\theta$$

14.21

$$\tan\theta\left[\frac{1}{1+\csc\theta}+\frac{1}{1+\sin\theta}\right] \equiv \tan\theta\left[\frac{1}{1+\frac{1}{\sin\theta}}+\frac{1}{1+\sin\theta}\right]$$

$$\equiv \tan\theta\left[\frac{\sin\theta}{1+\sin\theta}+\frac{1}{1+\sin\theta}\right] \equiv \tan\theta\left[\frac{\sin\theta+1}{1+\sin\theta}\right] \equiv \tan\theta$$

14.22 TRIG. IDENTITY

Prove the identity

$$\sec\theta(\sin^2\theta + \sin\theta - 1) \equiv \tan\theta - \cos\theta$$

14.22

$$\sec\theta(\sin^2\theta + \sin\theta - 1) \equiv \frac{1}{\cos\theta}(1 - \cos^2\theta + \sin\theta - 1)$$

$$\equiv \frac{1}{\cos\theta}(\sin\theta - \cos^2\theta) \equiv \tan\theta - \cos\theta.$$

TRIG. IDENTITY

Prove the identity $\frac{\cos\theta}{1-\sin\theta} \equiv \tan\theta + \sec\theta$

14.23

$$\frac{\cos\theta}{1-\sin\theta}\cdot 1 \equiv \frac{\cos\theta}{1-\sin\theta}\cdot\frac{1+\sin\theta}{1+\sin\theta} \equiv \frac{\cos\theta(1+\sin\theta)}{1-\sin^2\theta}$$

$$\equiv \frac{\cos\theta(1+\sin\theta)}{\cos^2\theta} \equiv \frac{1+\sin\theta}{\cos\theta} \equiv \sec\theta + \tan\theta$$

14.24 TRIG. IDENTITY

Prove the identity $\dfrac{1}{\sec\theta - \tan\theta} \equiv \sec\theta + \tan\theta$

14.24

$$\frac{1}{\sec\theta - \tan\theta} \cdot 1 = \frac{1}{\sec\theta - \tan\theta} \cdot (\sec^2\theta - \tan^2\theta)$$

$$= \sec\theta + \tan\theta.$$

Prove the identity $\frac{\sec^2\theta - 2}{\csc^2\theta - 2} = -\tan^2\theta$

14.25

$$\frac{\sec^2\theta - 2}{\csc^2\theta - 2} \equiv \frac{1+\tan^2\theta - 2}{1+\cot^2\theta - 2} \equiv \frac{\tan^2\theta - 1}{\cot^2\theta - 1} \equiv \frac{\tan^2\theta - 1}{\dfrac{1}{\tan^2\theta} - 1} \equiv \frac{\tan^2\theta - 1}{\dfrac{1-\tan^2\theta}{\tan^2\theta}}$$

$$\equiv \left(\tan^2\theta - 1\right)\left(\frac{\tan^2\theta}{1-\tan^2\theta}\right) \equiv -\tan^2\theta\,.$$

14.26

IDENTITIES: USE OF

Simplify $\dfrac{\sin\beta + \tan\beta}{\cot\beta + \csc\beta}$

14.26

$$\frac{\sin\beta + \tan\beta}{\cot\beta + \csc\beta} \equiv \frac{\sin\beta + \dfrac{\sin\beta}{\cos\beta}}{\dfrac{\cos\beta}{\sin\beta} + \dfrac{1}{\sin\beta}} \equiv \frac{\sin\beta(\cos\beta + 1)}{\cos\beta} \cdot \frac{\sin\beta}{\cos\beta + 1}$$

$$\equiv \frac{\sin^2\beta}{\cos\beta} \equiv \sin\beta\tan\beta$$

14.27 TRIG. IDENTITY: USE OF

If $x = r \sin \Theta \cos \Phi$

$y = r \sin \Theta \sin \Phi$

$z = r \cos \Theta$, show that $x^2 + y^2 + z^2 = r^2$

14.27

$$
\begin{aligned}
x^2 + y^2 + z^2 &= r^2 \sin^2\theta \cos^2\phi + r^2 \sin^2\theta \sin^2\phi + r^2 \cos^2\theta \\
&= r^2\left[\sin^2\theta(\cos^2\phi + \sin^2\phi) + \cos^2\theta\right] \\
&= r^2\left[\sin^2\theta(1) + \cos^2\theta\right] \\
&= r^2
\end{aligned}
$$

14.28 TRIG. IDENTITY: USE OF

Express $\frac{1+\cot^2\Phi}{\cot^2\ \Phi}$ in terms of $\cos\Phi$

14.28

$$\frac{1+\cot^2\phi}{\cot^2\phi} \equiv \frac{\csc^2\phi}{\cos^2\phi/\sin^2\phi} \equiv \frac{1/\sin^2\phi}{\cos^2\phi/\sin^2\phi} \equiv \frac{1}{\cos^2\phi}$$

14.29 TRIG. IDENTITY

Prove the identity

$$\frac{\tan A}{1+\sec A} - \frac{\tan A}{1-\sec A} \equiv 2\csc A$$

14.29

$$\frac{\tan A}{1+\sec A} - \frac{\tan A}{1-\sec A} \equiv \frac{\tan A - \sec A \tan A - \tan A - \sec A \tan A}{1-\sec^2 A}$$

$$\equiv \frac{-2\sec A \tan A}{-\tan^2 A} \equiv \frac{2\sec A}{\tan A} \equiv \frac{2}{\sin A} \equiv 2\csc A$$

14.30 IDENTITIES: USE OF

Express $\frac{\cos^2\theta - \sin^2\theta}{\csc^2\theta}$ in terms of $\sin\theta$

14.30

$$\frac{\cos^2\theta - \sin^2\theta}{\csc^2\theta} \equiv \frac{(1 - \sin^2\theta) - \sin^2\theta}{1/\sin^2\theta} \equiv (1 - 2\sin^2\theta)\sin^2\theta$$

15.1 INVERSE TRIG. FUNCTIONS

What is meant by $\theta = \sin^{-1} x$ or $\theta = \arcsin x$?

15.1

If one has the equation $x = \sin\theta$, and wishes to solve for θ, it is expressed in the form $\theta = \sin^{-1} x$ or $\theta = \arcsin x$. This is read as "θ is the angle whose sine is x."

Warning: $\sin^{-1} x$ is not the same as $(\sin x)^{-1}$.

15.2 INVERSE TRIG. FUNCTIONS

What are such functions as $\sin^{-1} x$, $\cos^{-1} x$, $\tan^{-1} x$, etc., called?

15.2

These are called the inverse trig. functions, and always are expressed as: the angle whose sine is x, or the angle whose cosine is x, or the angle whose tangent is x, etc.

What are the principal values of the inverse sine?

15.3

As $\theta = \sin^{-1}x$ is multiple-valued (has an unlimited number of solutions), the principal values are defined so that the inverse function becomes single-valued. The notation used will be: $\text{Sin}^{-1}x$ or Arc sin x. $-\frac{\pi}{2} \leq \text{Sin}^{-1}x \leq \frac{\pi}{2}$.

What are the principal values of the inverse cosine?

15.4

$$0 \leq \text{Cos}^{-1} x \leq \pi$$

What are the principal values of the inverse tangent?

15.5

$$-\frac{\pi}{2} < \mathrm{Tan}^{-1} x < \frac{\pi}{2}$$

15.6 INVERSE TRIG. FUNCTIONS

Evaluate $\sin \mathrm{Cos}^{-1} \frac{1}{2}$

15.6

$$\sin \mathrm{Cos}^{-1}\frac{1}{2} = \sin\frac{\pi}{3} = \frac{\sqrt{3}}{2}$$

Evaluate $\tan \text{Sin}^{-1}\left(-\frac{\sqrt{3}}{2}\right)$

15.7

$$\tan \mathrm{Sin}^{-1}\left(-\frac{\sqrt{3}}{2}\right) = \tan\left(-\frac{\pi}{3}\right) = -\tan\frac{\pi}{3} = -\sqrt{3}$$

15.8 INVERSE TRIG. FUNCTIONS

Evaluate $\sin\left(2\,\text{Sin}^{-1}\frac{1}{2}\right)$

15.8

$$\sin\left(2\mathrm{Sin}^{-1}\frac{1}{2}\right) = \sin\left(2\frac{\pi}{6}\right) = \sin\frac{\pi}{3} = \frac{\sqrt{3}}{2}$$

15.9 INVERSE TRIG. FUNCTIONS

Evaluate $\sec \operatorname{Cos}^{-1}\left(-\frac{1}{2}\right)$

15.9

$$\sec \operatorname{Cos}^{-1}\left(-\frac{1}{2}\right) = \sec\left(\frac{2\pi}{3}\right) = -\sec\frac{\pi}{3} = -2.$$

15.10 INVERSE TRIG. FUNCTIONS

Evaluate $\text{Sin}^{-1}(-1) - \text{Cos}^{-1}(-1)$

15.10

$$\mathrm{Sin}^{-1}(-1) - \mathrm{Cos}^{-1}(-1) = -\frac{\pi}{2} - (\pi) = -\frac{3\pi}{2}.$$

INVERSE TRIG. FUNCTIONS

Evaluate $\text{Tan}^{-1}\left(\frac{1}{\sqrt{3}}\right) - \frac{1}{2}\text{Cos}^{-1}\left(-\frac{1}{\sqrt{2}}\right)$

15.11

$$\mathrm{Tan}^{-1}\left(\frac{1}{\sqrt{3}}\right) - \frac{1}{2}\mathrm{Cos}^{-1}\ -\left(\frac{1}{\sqrt{2}}\right) = \frac{\pi}{6} - \frac{1}{2}\left(\frac{3\pi}{4}\right) = \frac{\pi}{6} - \frac{3\pi}{8} = -\frac{5\pi}{24}$$

INVERSE TRIG. FUNCTIONS

Evaluate $\sin\left(\operatorname{Sin}^{-1}\frac{1}{3} + \operatorname{Sin}^{-1}\frac{1}{5}\right)$

15.12

Let $A = \text{Sin}^{-1}\frac{1}{3}$ $\qquad$ $B = \text{Sin}^{-1}\frac{1}{5}$

then $\sin(A+B) = \sin A \cos B + \cos A \sin B = \frac{1}{3}\frac{\sqrt{24}}{5} + \frac{\sqrt{8}}{3}\frac{1}{5}$

$$= \frac{\sqrt{24} + \sqrt{8}}{15} = \frac{2\sqrt{6} + 2\sqrt{2}}{15} = \frac{2}{15}(\sqrt{6} + \sqrt{2})$$

15.13 INVERSE TRIG. FUNCTIONS

Evaluate $\mathrm{Cos}^{-1}\left(-\frac{1}{\sqrt{5}}\right) - \mathrm{Sin}^{-1}\left(-\frac{1}{\sqrt{10}}\right)$

15.13

Let $A = \text{Cos}^{-1}\left(-\frac{1}{\sqrt{5}}\right)$, A in Q II; $B = \text{Sin}^{-1}\left(-\frac{1}{\sqrt{10}}\right)$, B in Q IV

$$\cos(A-B) = \cos A \cos B + \sin A \sin B = -\frac{1}{\sqrt{5}}\ \frac{3}{\sqrt{10}} + \frac{2}{\sqrt{5}}\left(-\frac{1}{\sqrt{10}}\right)$$

$$= -\frac{5}{\sqrt{50}} = -\frac{1}{\sqrt{2}} \qquad \therefore A - B = \frac{3\pi}{4}$$

Evaluate $\tan\left[\text{Tan}^{-1} 3 - \text{Cot}^{-1} 2\right]$

15.14

Let $A = \mathrm{Tan}^{-1} 3$; $B = \mathrm{Cot}^{-1} 2$

$$\tan(A-B) \equiv \frac{\tan A - \tan B}{1+\tan A \tan B} = \frac{3 - \frac{1}{2}}{1+3\cdot\frac{1}{2}} = \frac{\frac{5}{2}}{\frac{5}{2}} = 1$$

15.15 INVERSE TRIG. FUNCTIONS

Find $\sin\left[\frac{1}{2}\mathrm{Cos}^{-1}\frac{4}{5}\right]$

15.15

Let $\cos^{-1}\frac{4}{5} = A$ (Quad. I)

$$\sin\frac{1}{2}A = +\sqrt{\frac{1 - \cos A}{2}} = \sqrt{\frac{1 - 4/5}{2}} = \sqrt{\frac{1}{10}}$$

Find x if $\sin\left[2\sin^{-1}x\right] = 1$.

15.16

Let $\text{Sin}^{-1} x = A$

$\sin 2A = 1$; $2A = \text{Sin}^{-1} 1 = \pi/2$; $A = \pi/4$

Then $\text{Sin}^{-1} x = \pi/4$; $x = \sin \pi/4 = \sqrt{2}/2$

16.1 IDENTITY: SUM OF TWO ANGLES

$$\sin(A+B) = ?$$

16.1

$$\sin(A+B) \equiv \sin A \cos B + \cos A \sin B$$

16.2 IDENTITY: DIFFERENCE OF TWO ANGLES

$$\sin (A-B) \equiv ?$$

16.2

$$\sin(A-B) \equiv \sin A \cos B - \cos A \sin B$$

16.3 IDENTITY: SUM OF TWO ANGLES

$$\cos(A + B) \equiv ?$$

16.3

$$\cos(A + B) \equiv \cos A \cos B - \sin A \sin B.$$

16.4 IDENTITY: DIFFERENCE OF TWO ANGLES

$$\cos (A - B) \equiv ?$$

16.4

$$\cos(A-B) \equiv \cos A \cos B + \sin A \sin B.$$

16.5 IDENTITY: SUM OF TWO ANGLES

$$\tan (A + B) \equiv ?$$

16.5

$$\tan(A+B) \equiv \frac{\tan A + \tan B}{1 - \tan A \tan B}$$

16.6 IDENTITY: DIFF. OF TWO ANGLES

$$\text{Tan}\ (A-B) \equiv ?$$

16.6

$$\tan(A-B) \equiv \frac{\tan A - \tan B}{1 + \tan A \tan B}$$

16.7 IDENTITY: DOUBLE ANGLE

$$\sin 2A \equiv ?$$

16.7

$$\sin 2A \equiv 2 \sin A \cos A$$

16.8 IDENTITY: DOUBLE ANGLE

$$\cos 2A \equiv ?$$

16.8

$$\cos 2A \equiv \cos^2 A - \sin^2 A \equiv 1 - 2\sin^2 A \equiv 2\cos^2 A - 1$$

16.9 IDENTITY: DOUBLE ANGLE

$$\tan 2A \equiv ?$$

16.9

$$\tan 2A \equiv \frac{2\tan A}{1-\tan^2 A}$$

16.10 IDENTITY: HALF ANGLE

$$\sin\frac{1}{2}A \equiv ?$$

16.10

$$\sin\frac{1}{2}A \equiv \pm\sqrt{\frac{1-\cos A}{2}}$$

Note: The sign used depends on the quadrant in which $\frac{1}{2}A$ lies.

16.11 IDENTITY: HALF ANGLE

$$\cos \frac{1}{2} A \equiv ?$$

16.11

$$\cos \frac{1}{2} A \equiv \pm \sqrt{\frac{1 + \cos A}{2}}$$

Note: The sign used depends on the quadrant in which $\frac{1}{2}A$ is located.

16.12 IDENTITY: HALF ANGLE

$$\tan \frac{1}{2} A \equiv ?$$

16.12

$$\tan\frac{1}{2}A \equiv \pm\sqrt{\frac{1-\cos A}{1+\cos A}} \equiv \frac{+\sin A}{1+\cos A} \equiv \frac{1-\cos A}{\sin A}$$

Note: The sign used in the first case depends upon the quadrant in which $\frac{1}{2}A$ lies.

16.13 IDENTITY: SUM TO PRODUCT

$\sin A + \sin B = ?$ (to convert a sum to a product.)

16.13

$$\sin A + \sin B \equiv 2 \sin \tfrac{1}{2}(A + B) \cdot \cos \tfrac{1}{2}(A - B)$$

16.14 IDENTITY: DIFFERENCE TO PRODUCT

$\sin A - \sin B = ?$ (to convert a difference to a product).

16.14

$$\sin A - \sin B \equiv 2\cos\frac{1}{2}(A+B)\sin\frac{1}{2}(A-B)$$

16.15 IDENTITY: SUM TO PRODUCT

$\cos A + \cos B \equiv$? (to convert a sum to a product).

16.15

$$\cos A + \cos B \equiv 2 \cos \frac{1}{2}(A+B) \cos \frac{1}{2}(A-B)$$

16.16 IDENTITY: DIFFERENCE TO PRODUCT

$\text{Cos } A - \text{Cos } B \equiv ?$ (to convert a difference to a product).

16.16

$$\cos A - \cos B \equiv -2 \sin\frac{1}{2}(A+B) \sin\frac{1}{2}(A-B)$$

16.17 IDENTITY: PRODUCT TO SUM

Sin A cos B ≡ ? (To convert a product to a sum)

16.17

$$\sin A \cos B \equiv \frac{1}{2}\left[\sin(A+B) + \sin(A-B)\right]$$

16.18 IDENTITY: PRODUCT TO SUM

$\text{Cos } A \cos B \equiv ?$ (to convert a product to a sum)

16.18

$$\cos A \cos B \equiv \frac{1}{2}\left[\cos(A+B) + \cos(A-B)\right]$$

16.19 IDENTITY: PRODUCT TO SUM

Sin A sin B ≡ ? (to convert a product to a sum.)

16.19

$$\sin A \sin B \equiv -\frac{1}{2}\left[\cos(A+B) - \cos(A-B)\right]$$

16.20 IDENTITY: SUM OF TWO ANGLES

Given $\sin A = -\frac{5}{13}$ (A in Q III) and $\cos B = -\frac{5}{13}$ (B in Q II)

find $\sin(A+B)$

16.20

$$\sin(A+B) \equiv \sin A \cos B + \cos A \sin B$$

$$= \left(\frac{-5}{13}\right)\left(\frac{-5}{13}\right) + \left(-\frac{12}{13}\right)\left(\frac{12}{13}\right) = \frac{25}{169} - \frac{144}{169} = -\frac{119}{169}$$

16.21 IDENTITIES: USE OF

Simplify $\sin 293^\circ \cos 83^\circ + \cos 293^\circ \sin 83^\circ$

16.21

$$\sin 293^\circ \cos 83^\circ + \cos 293^\circ \sin 83^\circ \equiv \sin(293^\circ + 83^\circ) \equiv \sin(376^\circ)$$
$$\equiv \sin 16^\circ$$

Find $\sin 195°$ without tables.

16.22

$$\sin 195^\circ = -\sin 15^\circ = -\sin(45^\circ - 30^\circ) =$$

$$-\left[\sin 45^\circ \cos 30^\circ - \cos 45^\circ \sin 30^\circ\right]$$

$$= -\left[\frac{1}{\sqrt{2}}\,\frac{\sqrt{3}}{2} - \frac{1}{\sqrt{2}}\cdot\frac{1}{2}\right] = -\left[\frac{\sqrt{3}-1}{2\sqrt{2}}\right] = \frac{1-\sqrt{3}}{2\sqrt{2}}$$

16.23 IDENTITIES: USE OF

Prove $\sin 3\theta \equiv 3 \sin\theta - 4 \sin^3 \theta$

16.23

$$\sin 3\theta \equiv \sin(\theta + 2\theta) \equiv \sin\theta\cos 2\theta + \cos\theta\sin 2\theta$$

$$\equiv \sin\theta(1 - 2\sin^2\theta) + \cos\theta(2\sin\theta\cos\theta)$$

$$\equiv \sin\theta(1 - 2\sin^2\theta) + 2\sin\theta(1 - \sin^2\theta)$$

$$\equiv 3\sin\theta - 4\sin^3\theta.$$

16.24

IDENTITIES: USE OF

Given $\sin A = -\frac{5}{13}$ (A in Q III) and $\cos B = -\frac{5}{13}$ (B in Q II) find $\sin(A-B)$.

16.24

$$\sin(A-B) \equiv \sin A \cos B - \cos A \sin B = \left(-\frac{5}{13}\right)\left(-\frac{5}{13}\right) - \left(-\frac{12}{13}\right)\left(\frac{12}{13}\right)$$

$$= \frac{25}{169} + \frac{144}{169} = 1$$

16.25 IDENTITY: SUM OF TWO ANGLES

Given $\sin A = -\frac{5}{13}$ (A in Q III) and $\cos B = -\frac{5}{13}$ (B in Q II) find $\cos(A+B)$

16.25

$$\cos(A+B) \equiv \cos A \cos B - \sin A \sin B$$
$$= \left(-\frac{12}{13}\right)\left(-\frac{5}{13}\right) - \left(-\frac{5}{13}\right)\left(\frac{12}{13}\right) = \frac{60}{169} + \frac{60}{169} = \frac{120}{169}$$

16.26 IDENTITY: USE OF

Simplify $\cos 23^\circ \cos 67^\circ - \sin 23^\circ \sin 67^\circ$

16.26

$$\cos 23^\circ \cos 67^\circ - \sin 23^\circ \sin 67^\circ \equiv \cos(23^\circ + 67^\circ) \equiv \cos 90^\circ = 0.$$

16.27 IDENTITIES: USE OF

Simplify $\cos 50^\circ \cos (50^\circ - x) + \sin 50^\circ \sin (50^\circ - x)$.

16.27

$$\cos 50^\circ \cos(50^\circ - x) + \sin 50^\circ \sin(50^\circ - x) \equiv \cos\left[50^\circ - (50^\circ - x)\right]$$
$$\equiv \cos x$$

Find $\cos 75°$ without tables.

16.28

$$\cos 75^\circ = \cos(45^\circ + 30^\circ) = \cos 45^\circ \cos 30^\circ - \sin 45^\circ \sin 30^\circ$$

$$= \frac{1}{\sqrt{2}} \cdot \frac{\sqrt{3}}{2} - \frac{1}{\sqrt{2}} \cdot \frac{1}{2} = \frac{\sqrt{3} - 1}{2\sqrt{2}} = \frac{\sqrt{6} - \sqrt{2}}{4}$$

16.29 IDENTITY: SUM OF TWO ANGLES

Derive a formula for $\cot(A + B)$ in terms of $\cot A$ and $\cot B$.

16.29

$$\cot(A+B) \equiv \frac{1}{\tan(A+B)} \equiv \frac{1 - \tan A \tan B}{\tan A + \tan B} \equiv \frac{1 - \frac{1}{\cot A} \cdot \frac{1}{\cot B}}{\frac{1}{\cot A} + \frac{1}{\cot B}}$$

$$\equiv \frac{\frac{\cot A \cot B - 1}{\cot A \cot B}}{\frac{\cot B + \cot A}{\cot A \cot B}} \equiv \frac{-1 + \cot A \cot B}{\cot A + \cot B}$$

16.30 IDENTITY: USE OF

Simplify $\dfrac{\tan 75^\circ - \tan 15^\circ}{1 + \tan 75^\circ \tan 15^\circ}$

16.30

$$\frac{\tan 75^\circ - \tan 15^\circ}{1 + \tan 75^\circ \tan 15^\circ} \equiv \tan(75^\circ - 15^\circ) \equiv \tan 60^\circ \equiv \sqrt{3}\,.$$

16.31 IDENTITY: SUM OF TWO ANGLES

Given $\sin A = -\frac{3}{5}$, (A in Q III), and $\sin B = \frac{8}{17}$, (B in Q II) find $\tan(A+B)$.

16.31

$$\tan(A+B) \equiv \frac{\tan A + \tan B}{1 - \tan A \tan B} \equiv \frac{3/4 + (-8/15)}{1-(3/4)(-8/15)} \equiv \frac{\frac{45-32}{60}}{\frac{60+24}{60}} \equiv \frac{13}{84}$$

16.32 IDENTITY: DIFFERENCE OF TWO ANGLES

Given $\sin A = -\frac{3}{5}$ (A in QIII) and $\sin B = \frac{8}{17}$ (B in QII) find $\tan(A - B)$.

16.32

$$\tan(A-B) \equiv \frac{\tan A - \tan B}{1 + \tan A \tan B} \equiv \frac{3/4 - (-8/15)}{1 + (3/4)(-8/15)}$$

$$\equiv \frac{\dfrac{45+32}{60}}{\dfrac{60-24}{60}} \equiv \frac{77}{36}$$

16.33 IDENTITIES: USE OF

Find tan 165° without tables.

16.33 $\tan 165^\circ \equiv -\tan 15^\circ \equiv -\tan(45^\circ - 30^\circ) \equiv \tan(30^\circ - 45^\circ) \equiv$

$$\frac{\tan 30^\circ - \tan 45^\circ}{1 + \tan 30^\circ \cdot \tan 45^\circ} \equiv \frac{\frac{1}{\sqrt{3}} - 1}{1 + \frac{1}{\sqrt{3}} \cdot 1} \equiv \frac{\frac{(1 - \sqrt{3})}{\sqrt{3}}}{\frac{(1 + \sqrt{3})}{\sqrt{3}}} \equiv \frac{1 - \sqrt{3}}{1 + \sqrt{3}} \cdot \frac{1 - \sqrt{3}}{1 - \sqrt{3}}$$

$$\equiv \frac{1 - 2\sqrt{3} + 3}{1 - 3} \equiv \frac{4 - 2\sqrt{3}}{-2} \equiv \sqrt{3} - 2$$

16.34 IDENTITY: DOUBLE ANGLE

Given $\sin\theta = -\frac{3}{5}$ (θ in QIV)

find $\sin 2\theta$

16.34

$$\sin 2\theta \equiv 2\sin\theta\cos\theta \equiv 2\cdot\left(-\frac{3}{5}\right)\left(+\frac{4}{5}\right) \equiv -\frac{24}{25}$$

16.35 IDENTITY: DOUBLE ANGLE

Given $\sin\theta = -3/5$ (θ in Q IV) find $\cos 2\theta$.

16.35

$$\cos 2\theta \equiv \cos^2\theta - \sin^2\theta \equiv 1 - 2\sin^2\theta \equiv 1 - 2\left(\frac{9}{25}\right) \equiv +\frac{7}{25}$$

16.36 IDENTITY: DOUBLE ANGLE

Given $\sin\theta = -\frac{3}{5}$ (θ in Q IV) find $\tan 2\theta$.

16.36

$$\tan 2\Theta \equiv \frac{2\tan\Theta}{1-\tan^2\Theta} \equiv \frac{2\left(-\frac{3}{4}\right)}{1-\left(-\frac{3}{4}\right)^2} \equiv \frac{-\frac{3}{2}}{1-\frac{9}{16}} \equiv \frac{-\frac{3}{2}}{\frac{7}{16}} \equiv -\frac{3}{2}\left(\frac{16}{7}\right) \equiv -\frac{24}{7}$$

16.37 TRIG. IDENTITY

Prove the identity $\frac{1-\cos 2\theta}{\sin 2\theta} \equiv \tan\theta$

16.37

$$\frac{1-\cos 2\theta}{\sin 2\theta} \equiv \frac{1-(1-2\sin^2\theta)}{2\sin\theta\cos\theta}$$

$$\equiv \frac{2\sin^2\theta}{2\sin\theta\cos\theta}$$

$$\equiv \frac{\sin\theta}{\cos\theta} \equiv \tan\theta$$

TRIG. IDENTITY

Prove the identity $\frac{\sin 5\theta}{\sin\theta} - \frac{\cos 5\theta}{\cos\theta} \equiv 4\cos 2\theta$

16.38

$$\frac{\sin 5\theta}{\sin\theta} - \frac{\cos 5\theta}{\cos\theta} \equiv \frac{\sin 5\theta \cos\theta - \cos 5\theta \sin\theta}{\sin\theta\cos\theta} \equiv \frac{\sin(5\theta-\theta)}{\sin\theta\cos\theta}$$

$$\equiv \frac{\sin 4\theta}{\sin\theta\cos\theta} \equiv \frac{2\sin 2\theta\cos 2\theta}{\frac{1}{2}\sin 2\theta} \equiv 4\cos 2\theta.$$

Given $\cos\theta = \frac{4}{5}$ (θ in Q I)

find $\sin\frac{1}{2}\theta$ and $\cos\frac{1}{2}\theta$ by first finding $\tan\frac{1}{2}\theta$.

16.39

$$\tan\frac{1}{2}\theta \equiv \frac{1-\cos\theta}{\sin\theta} \equiv \frac{1 - 4/5}{3/5} = \frac{1}{3}$$

$$\therefore \sin\frac{1}{2}\theta = \frac{1}{\sqrt{10}} \; ; \qquad \cos\frac{1}{2}\theta = \frac{3}{\sqrt{10}}$$

16.40 IDENTITY: HALF ANGLE

Given $\sin\theta = -\frac{24}{25}$ (θ in Q III) find $\sin\frac{\theta}{2}$ directly.

16.40

$$\sin\theta = -\frac{24}{25}; \qquad \cos\theta = -\frac{7}{25}$$

As $180° < \theta < 270°$, $90° < \frac{\theta}{2} < 135°$ $\left(\frac{\theta}{2} \text{ in Q II}\right)$

$$\therefore \sin\frac{\theta}{2} \equiv +\sqrt{\frac{1-\cos\theta}{2}} \equiv +\sqrt{\frac{1+7/25}{2}} \equiv \sqrt{\frac{32/25}{2}} = \sqrt{\frac{16}{25}} = \frac{4}{5}$$

16.41 IDENTITY: USE OF

Prove the identity $\sin\theta \tan\frac{1}{2}\theta + 2\cos\theta = 2\cos^2\frac{1}{2}\theta$

16.41

$$\sin\theta \tan\frac{1}{2}\theta + 2\cos\theta \equiv \sin\theta\left(\frac{1-\cos\theta}{\sin\theta}\right) + 2\cos\theta \equiv 1 - \cos\theta + 2\cos\theta$$

$$\equiv 1 + 2\cos\theta \equiv 2\left(\frac{1+\cos\theta}{2}\right) \equiv 2\cos^2\frac{1}{2}\theta.$$

16.42 TRIG. IDENTITY

Simplify $\sin 75^\circ \cos 75^\circ$

16.42

$$\sin 75^\circ \cos 75^\circ \equiv \frac{1}{2}(2 \sin 75^\circ \cos 75^\circ) \equiv \frac{1}{2} \sin 150^\circ$$

$$\equiv \frac{1}{2} \sin 30^\circ \equiv \frac{1}{2}\left(\frac{1}{2}\right) \equiv \frac{1}{4}$$

16.43 IDENTITY - SUM TO PRODUCT

Express the following sum as a product, and evaluate.

$$\sin 75^\circ + \sin 15^\circ$$

16.43

$$\sin 75^\circ + \sin 15^\circ \equiv 2 \sin \frac{1}{2}(75^\circ + 15^\circ) \cos \frac{1}{2}(75^\circ - 15^\circ)$$

$$\equiv 2 \sin \frac{1}{2}(90^\circ) \cos \frac{1}{2}(60^\circ) \equiv 2 \sin 45^\circ \cos 30^\circ$$

$$\equiv 2 \frac{1}{\sqrt{2}} \frac{\sqrt{3}}{2} \equiv \frac{\sqrt{3}}{\sqrt{2}} \equiv \frac{\sqrt{6}}{2}.$$

Express the following difference as a product.

$$\sin 100^\circ - \sin 50^\circ$$

16.44

$$\sin 100^\circ - \sin 50^\circ \equiv 2\cos\frac{1}{2}(100^\circ + 50^\circ)\sin\frac{1}{2}(100^\circ - 50^\circ)$$

$$= 2\cos 75^\circ \sin 25^\circ$$

16.45 IDENTITIES: USE OF

Evaluate $\cos 20^\circ + \cos 140^\circ + \cos 260^\circ$.

16.45

$$\cos 20^\circ + \cos 140^\circ \equiv 2\cos\frac{1}{2}(20^\circ + 140^\circ)\ \cos\frac{1}{2}(20^\circ - 140^\circ)$$

$$\equiv 2\cos 80^\circ \cos(-60^\circ) \equiv 2\cos 80^\circ \cos 60^\circ$$

$$\cos 260^\circ \equiv -\cos 80^\circ$$

$$\therefore \cos 20^\circ + \cos 140^\circ + \cos 260^\circ \equiv 2\cos 80^\circ \cos 60^\circ - \cos 80^\circ$$

$$\equiv \cos 80^\circ(2\cos 60^\circ - 1) \quad \equiv \cos 80^\circ\left(2\cdot\frac{1}{2} - 1\right) \equiv 0$$

16.46 IDENTITY: DIFF. TO PRODUCT

Express the following difference as a product, and simplify

$$\cos\frac{5\pi}{8} - \cos\frac{3\pi}{8}$$

16.46

$$\cos\frac{5\pi}{8} - \cos\frac{3\pi}{8} \equiv -2\sin\frac{1}{2}\left(\frac{5\pi}{8}+\frac{3\pi}{8}\right)\cdot \sin\frac{1}{2}\left(\frac{5\pi}{8}-\frac{3\pi}{8}\right)$$

$$\equiv -2\sin\frac{\pi}{2}\sin\frac{\pi}{8} \equiv -2\cdot 1\cdot \sin\frac{\pi}{8} \equiv -2\sin\frac{\pi}{8}$$

16.47 IDENTITY: PRODUCT TO SUM

Express the following product as a sum, and evaluate.

$$\sin 75^\circ \cos 15^\circ$$

16.47

$$\sin 75^\circ \cos 15^\circ \equiv \frac{1}{2}\left[\sin(75^\circ + 15^\circ) + \sin(75^\circ - 15^\circ)\right]$$

$$\equiv \frac{1}{2}\left[\sin 90^\circ + \sin 60^\circ\right]$$

$$\equiv \frac{1}{2}\left[1 + \frac{\sqrt{3}}{2}\right] \equiv \frac{2 + \sqrt{3}}{4}$$

16.48 IDENTITY: PRODUCT TO SUM

Express the following product as a sum, and evaluate.

$$\sin 105^\circ \sin 75^\circ$$

16.48

$$\sin 105^\circ \sin 75^\circ \equiv -\frac{1}{2}\left[\cos(105^\circ + 75^\circ) - \cos(105^\circ - 75^\circ)\right]$$

$$\equiv -\frac{1}{2}\left[\cos 180^\circ - \cos 30^\circ\right] \equiv -\frac{1}{2}\left[-1 - \frac{\sqrt{3}}{2}\right]$$

$$\equiv \frac{2 + \sqrt{3}}{4}$$

16.49 TRIG. IDENTITY

Express the following product as a sum, and evaluate.

$$\cos\frac{5\pi}{8}\cdot\cos\frac{\pi}{8}$$

16.49

$$\cos\frac{5\pi}{8}\cdot\cos\frac{\pi}{8} \equiv \frac{1}{2}\left[\cos\left(\frac{5\pi}{8}+\frac{\pi}{8}\right)+\cos\left(\frac{5\pi}{8}-\frac{\pi}{8}\right)\right] \equiv \frac{1}{2}\left[\cos\frac{3\pi}{4}+\cos\frac{\pi}{2}\right]$$

$$\equiv \frac{1}{2}\left[-\cos\frac{\pi}{4}+\cos\frac{\pi}{2}\right] \equiv \frac{1}{2}\left[-\frac{1}{\sqrt{2}}+0\right]$$

$$\equiv -\frac{1}{2\sqrt{2}} \equiv -\frac{\sqrt{2}}{4}$$

Prove the identity $\dfrac{\sin 2\beta + \sin\beta}{\cos 2\beta + 3\cos\beta + 2} \equiv \tan\dfrac{1}{2}\beta$

16.50

$$\frac{\sin 2\beta + \sin \beta}{\cos 2\beta + 3 \cos \beta + 2} \equiv \frac{2 \sin \beta \cos \beta + \sin \beta}{2 \cos^2 \beta - 1 + 3 \cos \beta + 2} \equiv$$

$$\frac{\sin \beta \cancel{(2 \cos \beta + 1)}}{\cancel{(2 \cos \beta + 1)}(\cos \beta + 1)} \equiv \frac{\sin \beta}{\cos \beta + 1} \equiv \tan \frac{1}{2} \beta .$$

Prove the identity

$$\frac{\cos\frac{1}{2}A + \sin\frac{1}{2}A}{\cos\frac{1}{2}A - \sin\frac{1}{2}A} = \tan A + \sec A$$

16.51 $$\frac{\cos\frac{1}{2}A+\sin\frac{1}{2}A}{\cos\frac{1}{2}A-\sin\frac{1}{2}A}\ \frac{\cos\frac{1}{2}A+\sin\frac{1}{2}A}{\cos\frac{1}{2}A+\sin\frac{1}{2}A} \equiv$$

$$\frac{\cos^2\frac{1}{2}A+\sin^2\frac{1}{2}A+2\cos\frac{1}{2}A\sin\frac{1}{2}A}{\cos^2\frac{1}{2}A-\sin^2\frac{1}{2}A} \equiv \frac{1+\sin A}{\cos A} \equiv \sec A+\tan A$$

17.1 TRIG. EQUATION

What is a trigonometric equation?

17.1

A trigonemetric equation is an equation, involving trigonometric functions, which is true only for particular values of the angles involved.

17.2 TRIG. EQUATION: PRIMARY SOLUTION

What is meant by the primary solutions of a trigonometric equation?

17.2

The primary solutions are those solutions in the range $0° \leq \theta < 360°$, although all angles coterminal with those could also be solutions of the equation.

Find the primary solutions of the equation $\tan\theta = -\frac{1}{\sqrt{3}}$.

17.3

The reference angle is the acute angle whose tangent is $1/\sqrt{3}$ or $30°$. As the tangent is negative in QII and IV, the solutions are $\theta = 150°$ and $330°$.

17.4 TRIG. EQUATION

Find the primary solution of $\cot \theta = 1$

17.4

The reference angle is 45°. As the cotangent is positive in QI and QIII, the solution is 45° and 225°.

17.5 TRIG. EQUATION

Find the primary solution of $\cos^2 \Phi = 4$.

17.5

$$\cos^2\theta = 4$$

$$\cos\theta = \pm 2$$

No solution, as the cosine of an angle cannot numerically exceed 1.

17.6 TRIG. EQUATION

Find the primary solution of $\sin\theta = \cos\theta$.

17.6

$$\sin\theta = \cos\theta$$

$$\frac{\sin\theta}{\cos\theta} = 1$$

$$\tan\theta = 1$$

$$\theta = 45^\circ, 225^\circ$$

17.7 TRIG. EQUATION

Find the primary solutions of the equation
$2\cos^2\theta - \cos\theta - 1 = 0$

17.7

$$2\cos^2\theta - \cos\theta - 1 = 0$$

$$(2\cos\theta + 1)(\cos\theta - 1) = 0$$

$$\cos\theta = -\frac{1}{2},\ 1$$

$$\theta = 120^\circ,\ 240^\circ,\ 0^\circ$$

17.8 TRIG. EQUATION

Find the primary solutions of $2\cos^2\theta + \sin\theta - 1 = 0$

17.8

$$2\cos^2\theta + \sin\theta - 1 = 0$$

$$2(1 - \sin^2\theta) + \sin\theta - 1 = 0$$

$$2\sin^2\theta - \sin\theta - 1 = 0$$

$$(2\sin\theta + 1)(\sin\theta - 1) = 0$$

$$\sin\theta = -\frac{1}{2},\ 1;\quad \theta = 210,\ 330°,\ 90°.$$

17.9 TRIG. EQUATION

Find the primary solutions of

$$\sin \theta - \cos \theta = 1$$

17.9

$$\sin\theta = \cos\theta + 1$$

$$+\sqrt{1-\cos^2\theta} = \cos\theta + 1$$

$$1 - \cos^2\theta = \cos^2\theta + 2\cos\theta + 1$$

$$2\cos^2\theta + 2\cos\theta = 0$$

$$2\cos\theta(\cos\theta + 1) = 0$$

$\cos\theta = 0, -1;\qquad \theta = 90°, 180°, 270°$

However, $\theta = 270°$ does not check the original equation.

$\therefore\ \theta = 90°, 180°$

Find the primary solutions of

$$\cos A \cot A - \sin A = \frac{3}{2} + \csc A$$

17.10 $$\cos A \cdot \frac{\cos A}{\sin A} - \sin A = \frac{3}{2} + \frac{1}{\sin A}$$

$$2\cos^2 A - 2\sin^2 A = 3\sin A + 2$$

$$2(1 - \sin^2 A) - 2\sin^2 A = 3\sin A + 2$$

$$4\sin^2 A + 3\sin A = 0$$

$$\sin A(4\sin A + 3) = 0$$

$$\sin A = 0,\ -3/4;\quad A = 0^\circ,\ 228^\circ.6,\ 311^\circ.4.$$

However, 0° does not check the original equation.

17.11 TRIG. EQUATION

Find the primary solution of $2\sin 3A - \sqrt{3}\tan 3A = 0$

17.11

$$2 \sin 3A - \sqrt{3}\left(\frac{\sin 3A}{\cos 3A}\right) = 0$$

$$\sin 3A\,(2 - \sqrt{3}\sec 3A) = 0$$

$$\sin 3A = 0; \qquad \sec 3A = 2/\sqrt{3}$$

$$3A = 0°, 180°, 360°, 540°, 720°, 900°;$$
$$30°, 330°, 390°, 690°, 750°, 1050°$$

$$A = 0°, 60°, 120°, 180°, 240°, 300°,$$
$$10°, 110°, 130°, 230°, 250°, 350°.$$

Find the primary solutions of

$$\csc^2\theta = 1 + \cot\theta$$

17.12

$$\csc^2\theta = 1 + \cot\theta$$

$$1 + \cot^2\theta = 1 + \cot\theta$$

$$\cot^2\theta - \cot\theta = 0$$

$$\cot\theta(\cot\theta - 1) = 0$$

$$\cot\theta = 0, 1; \quad \theta = 90°, 270°, 45°, 225°.$$

17.13 TRIG. EQUATION

Find the primary solutions of the equation $\cos(60^\circ + \theta)\cos(60^\circ - \theta) - \sin(60^\circ + \theta)\sin(60^\circ - \theta) = \tan\theta$.

17.13

$$\cos A \cos B - \sin A \sin B \equiv \cos(A+B)$$
$$\cos(60^\circ+\theta)\cos(60^\circ-\theta) - \sin(60^\circ+\theta)\sin(60^\circ-\theta) \equiv$$
$$\cos\ [(60^\circ+\theta)+(60^\circ-\theta)] \equiv \cos 120^\circ.$$

$$\therefore \tan\theta = -\frac{1}{2};\quad \theta = 153^\circ\!.4,\ 333^\circ\!.4$$

17.14 TRIG. EQUATION

Find the primary solutions of the equation

$$\cos 2\theta + \sin 2\theta = 1.$$

17.14

$$\cos 2\theta + \sin 2\theta = 1$$

$$1 - 2\sin^2\theta + 2\sin\theta\cos\theta = 1$$

$$2\sin\theta(\cos\theta - \sin\theta) = 0$$

$$\sin\theta = 0; \quad \cos\theta - \sin\theta = 0 \text{ or } \tan\theta = 1$$

$$\theta = 0^\circ, 180^\circ, 45^\circ, 225^\circ.$$

TRIG. EQUATION

Find the values of $\tan\theta$ that satisfy the equation $\tan 2\theta = \frac{4}{3}$.

17.15

Let $\tan\theta = x$

$$\tan 2\theta = \frac{2\tan\theta}{1-\tan^2\theta} = \frac{4}{3}$$

$$\frac{2x}{1-x^2} = \frac{4}{3}$$

$$6x = 4 - 4x^2$$

$$4x^2 + 6x - 4 = 0$$

$$2x^2 + 3x - 2 = 0$$

$$(2x-1)(x+2) = 0 \qquad x = \frac{1}{2},\ -2$$

Find the primary solutions of the equation sin

$$\sin^2\frac{1}{2}\theta - \sin\theta = 0$$

17.16

$$\sin^2\frac{1}{2}\theta - \sin\theta = 0$$

$$\frac{1-\cos\theta}{2} - \sin\theta = 0$$

$$\cos\theta = 1 - 2\sin\theta$$

$$\pm\sqrt{1-\sin^2\theta} = 1 - 2\sin\theta$$

$$1 - \sin^2\theta = 1 - 4\sin\theta + 4\sin^2\theta$$

$$5\sin^2\theta - 4\sin\theta = 0$$

$$\sin\theta(5\sin\theta - 4) = 0$$

$$\sin\theta = 0,\ 4/5$$

$$\theta = 0°,\ 180°,\ 53.°1,\ 126.°9$$

However, only 0° and 126.°9 satisfy the original equation.

17.17 TRIG. EQUATION

Find the primary solutions of the equation

$$\cos 3x + \cos x = 0.$$

17.17

$$\cos 3x + \cos x \equiv 2\cos\frac{1}{2}(3x+x)\cos\frac{1}{2}(3x-x) \equiv 2\cos 2x\cos x$$

$$\therefore 2\cos 2x\cos x = 0; \quad \cos 2x = 0; \quad \cos x = 0$$

$$x = \frac{90^\circ}{2}, \frac{270^\circ}{2}, \frac{450^\circ}{2}, \frac{630^\circ}{2}, 90^\circ, 270^\circ = 45^\circ, 135^\circ, 225^\circ, 315^\circ, 90^\circ, 270^\circ$$

17.18 TRIG. EQUATION

Find the primary solutions of the equation

$$\sin\theta \cos 4\theta - \sin 2\theta \cos 3\theta = 0.$$

17.18

$\sin\theta\cos 4\theta \equiv \frac{1}{2}\left[\sin 5\theta + \sin(-3\theta)\right]$

$\sin 2\theta\cos 3\theta \equiv \frac{1}{2}\left[\sin 5\theta + \sin(-\theta)\right]$

$\sin\theta\cos 4\theta - \sin 2\theta\cos 3\theta = 0$

$\frac{1}{2}(\sin 5\theta - \sin 3\theta) - \frac{1}{2}(\sin 5\theta - \sin\theta) = 0 \qquad \sin\theta - \sin 3\theta = 0$

$2\cos\frac{1}{2}(4\theta)\sin\frac{1}{2}(-2\theta) = 0$

$-\cos 2\theta\sin\theta = 0;\ \cos 2\theta = 0;\ \sin\theta = 0$

$\theta = \frac{90^\circ}{2}, \frac{270^\circ}{2}, \frac{450^\circ}{2}, \frac{630^\circ}{2}, 0^\circ, 180^\circ = 45^\circ, 135^\circ, 225^\circ, 0^\circ, 180^\circ$

Solve the simultaneous equations?

$$r(1+\cos\theta) = 3$$
$$r = 4(1-\cos\theta)$$

17.19 Solving both equations for r:

$$r = \frac{3}{1+\cos\theta} = 4(1-\cos\theta)$$

$$\frac{3}{4} = 1 - \cos^2\theta; \quad \cos^2\theta = \frac{1}{4}; \quad \cos\theta = \pm\frac{1}{2}$$

For $\cos\theta = \frac{1}{2}$, $r = 2$. For $\cos\theta = -\frac{1}{2}$, $r = 6$.

$\therefore$ Solutions for $0 \leq \theta \leq 360°$ are $r = 2$, $\theta = 60°$ or $300°$
$r = 6$, $\theta = 120°$ or $240°$

18.1 OBLIQUE TRIANGLE

What is an oblique triangle?

18.1

An oblique triangle is a triangle which does not contain a right angle.

18.2 OBLIQUE TRIANGLE SOLUTION

What is necessary to solve an oblique triangle?

18.2

Of the six parts, three have to be known in order to solve the triangle, including at least one side. There are four possibilities, known as Cases I, II, III, IV.

Case I: Two angles and a side.
Case II: Two sides and an angle opposite one.
Case III: Two sides and the included angle.
Case IV: Three sides.

18.3 OBLIQUE TRIANGLE SOLUTION: LAW OF SINES

What is the law of sines?

18.3

In any triangle the sides are proportional to the sines of the opposite angle.

$$\frac{a}{\sin A} = \frac{b}{\sin B} = \frac{c}{\sin C}$$

18.4 OBLIQUE TRIANGLES: LAW OF COSINES

What is the law of cosines?

18.4

In any triangle the square of any side equals the sum of the squares of the other two sides minus twice the product of these two sides and the cosine of their included angle. For example:

$$c^2 = a^2 + b^2 - 2ab \cos C \quad \text{or} \quad \cos C = \frac{a^2 + b^2 - c^2}{2ab}$$

18.5 OBLIQUE TRIANGLE SOLUTION: LAW OF TANGENTS

What is the law of tangents?

18.5

In any triangle the difference of any two sides is to their sum as the tangent of half the difference of the opposite angles is to the tangent of half the sum of these angles. For example,

$$\frac{a-b}{a+b}=\frac{\tan\frac{1}{2}(A-B)}{\tan\frac{1}{2}(A+B)}$$

18.6 OBLIQUE TRIANGLE SOLUTION: HALF ANGLE FORMULAS

What are the half-angle formulas?

18.6

If A is any angle of a triangle, $\tan\frac{1}{2}A = \frac{r}{s-a}$, where

$r = \sqrt{\frac{(s-a)(s-b)(s-c)}{s}}$ and $s = \frac{1}{2}(a+b+c)$

18.7 OBLIQUE TRIANGLE SOLUTION

How are triangles in Case I solved?

18.7

These triangles, where one side and two angles are known; can be solved completely by the law of sines.

How are triangles in Case II solved?

18.8

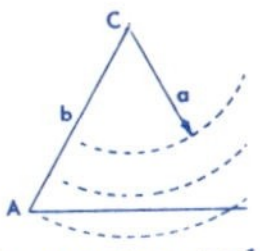

These triangles, where two sides and the angle opposite one of them are given, are ambiguous since 1, 0, or 2 solutions are possible. Suppose a,b and A are given. If A is constructed and b marked off on one side, from C draw an arc of length a. If a solution exists, it can be solved by the law of sines.

18.9 OBLIQUE TRIANGLE SOLUTION

How are triangles in Case III solved?

18.9

a) If the numbers are simple and only the third side is desired; by the law of cosines.
b) Otherwise by the law of tangents and the law of sines, as the law of cosines does not adapt itself to computation with logarithms.

18.10 OBLIQUE TRIANGLE SOLUTION

How are triangles in Case IV solved?

18.10

a) If the numbers are simple, by the law of cosines.
b) Otherwise by the half angle formulas.

What is an area formula for a case I triangle (two angles and a side)?

18.11

$$K = \frac{a^2 \sin B \sin C}{2 \sin A}$$ where a is the given side.

18.12 OBLIQUE TRIANGLE SOLUTION

What is an area formula for a Case III triangle (two sides and an included angle)?

18.12

$$K = \frac{1}{2} ab \sin C$$

18.13 OBLIQUE TRIANGLE SOLUTION: AREA

What is an area formula for a case IV triangle (three sides)?

18.13

$$K = rs = \sqrt{s(s-a)(s-b)(s-c)}$$

$$\text{where } s = \frac{1}{2}(a+b+c)$$

18.14 OBLIQUE TRIANGLE SOLUTION

What are good equations to use to check the solution of an oblique triangle?

18.14

Mollweide's Equations:

$$\frac{a-b}{c} = \frac{\sin\frac{1}{2}(A-B)}{\cos\frac{1}{2}C}$$

$$\frac{a+b}{c} = \frac{\cos\frac{1}{2}(A-B)}{\sin\frac{1}{2}C}$$

These are good check equations because each one con tains all six parts of the triangle.

18.15 OBLIQUE TRIANGLE SOLUTION

Show the solution of an oblique triangle given $A = 20^\circ$, $B = 50^\circ$ and $a = 10$.

18.15

$$C = 180° - (A+B) = 110°$$

$$\frac{10}{\sin 20°} = \frac{b}{\sin 50°} = \frac{c}{\sin 110°}$$

$$b = \frac{10 \sin 50°}{\sin 20°} \; ; \; c = \frac{10 \sin 110°}{\sin 20°} = \frac{10 \sin 70°}{\sin 20°} = 10 \cot 20°$$

18.16 OBLIQUE TRIANGLE SOLUTION

Show the solution of an oblique triangle given $A = 20^\circ$, $b = 5$ and $a = 2$.

18.16

$$\frac{2}{\sin 20^\circ} = \frac{5}{\sin B}$$

$$\sin B = \frac{5}{2} \sin 20^\circ$$

This will have two solutions; one case with B acute and the other with B obtuse.

18.17 OBLIQUE TRIANGLE SOLUTION

Show the solution of an oblique triangle given $a = 5$, $b = 6$ and $C = 60^\circ$.

18.17

$$c^2 = a^2 + b^2 - 2ab\cos C = 25 + 36 - 60\left(\frac{1}{2}\right) = 25 + 36 - 30 = 31$$

$$c = \sqrt{31}$$

Then $\dfrac{\sin A}{5} = \dfrac{\sin B}{6} = \dfrac{\sin 60^\circ}{\sqrt{31}}$; $A = 51^\circ$; $B = 69^\circ$

18.18 OBLIQUE TRIANGLE SOLUTION

Find angle A of an oblique triangle if $a = 3$, $b = 4$ and $c = 6$.

18.18

By the Law of Cosines,

$$\cos A = \frac{b^2 + c^2 - a^2}{2bc} = \frac{16 + 39 - 9}{48} = \frac{43}{48}$$

$$A = 26^{\circ}.3$$

18.19 OBLIQUE TRIANGLE SOLUTION

By the half-angle formulas, find angle A if $a = 3$, $b = 4$, $c = 6$.

18.19

$$s = \frac{3+4+6}{2} = 6.5$$

$$r = \sqrt{\frac{(3.5)(2.5)(.5)}{6.5}} = .82$$

$$\tan\frac{1}{2}A = \frac{.82}{3.5}; \qquad \frac{1}{2}A = 13^{\circ}.15; \qquad A = 26^{\circ}.3$$

OBLIQUE TRIANGLE SOLUTION

By the law of tangents, find A and B given $a = 5$, $b = 6$, and $C = 60°$.

18.20

$$\frac{b-a}{b+a}=\frac{\tan\frac{1}{2}(B-A)}{\tan\frac{1}{2}(B+A)}$$

$$\frac{1}{11}=\frac{\tan\frac{1}{2}(B-A)}{\tan 60^\circ}$$

$$\frac{1}{2}(B-A)= 9^\circ$$

$$\frac{1}{2}(B+A)=60^\circ$$

$$B=69^\circ$$

$$A=51^\circ$$

18.21 OBLIQUE TRIANGLE SOLUTION

Find the area of a triangle given $A=20^\circ$, $B=50^\circ$ and $a= 10$.

18.21

$$K = \frac{a^2 \sin B \sin C}{2 \sin A} = \frac{100 \cdot \sin 50^\circ \cdot \sin 110^\circ}{2 \cdot \sin 20^\circ} = \frac{50 \sin 50^\circ \cdot \sin 70^\circ}{\sin 20^\circ} = 105$$

18.22 OBLIQUE TRIANGLE SOLUTION

Find the area of a triangle given $a = 5$, $b = 6$, and $C = 60°$.

18.22

$$K = \frac{1}{2} a \cdot b \cdot \sin C = \frac{1}{2} \cdot 30 \cdot \sin 60° = 15 \cdot \frac{\sqrt{3}}{2} = 13.$$

18.23 OBLIQUE TRIANGLE SOLUTION

Find the area of a triangle given $a=3$, $b=4$, $c=6$.

18.23

$$K = \sqrt{s(s-a)(s-b)(s-c)}$$

$$s = \frac{1}{2}(a+b+c) = 6.5$$

$$K = \sqrt{(6.5)(3.5)(2.5)(.5)} = \frac{1}{4}\sqrt{455} = 5.33$$

18.24 OBLIQUE TRIANGLE SOLUTION

A tree stands erect on a slope that is inclined 8° from the horizontal. When the angle of elevation of the sun is 53° 30', the tree has a shadow 61 feet long directly down the slope. Find the height of the tree.

18.24

In the ΔABC, $\angle BAC = 45°30'$, $\angle ABC = 98°$, $AB = 61$ feet.

Find BC.

$$\frac{BC}{\sin\angle BAC} = \frac{61}{\sin\angle ACB}$$

$$BC = \frac{61 \sin 45° 30}{\sin 36° 30'} =$$

73.1 ft.

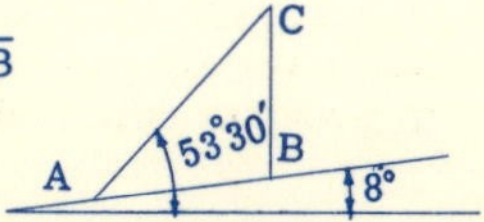

18.25 OBLIQUE TRIANGLE SOLUTION

In the triangle ABC, a, c, and A are given. If a=48 and c=60, find the values of A for which the triangle has:

a) no solution.

b) one solution.

c) two solutions.

18.25

Construct a right triangle with one leg 48 and hypotenuse 60.

Then there will be:
No solution if $A > 53° 7'$
One solution if $A = 53° 7'$
Two solutions if $A < 53° 7'$

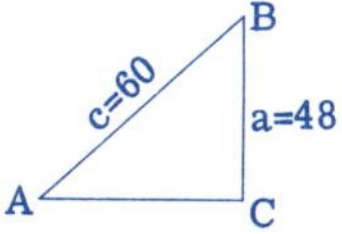

$\sin A = \frac{4}{5}$

$A = 53° 7'$

18.26 OBLIQUE TRIANGLE SOLUTION

Find the length of a lake if from a point 4 miles from one end and 5 miles from the other, the lake subtends an angle of $120°$.

18.26

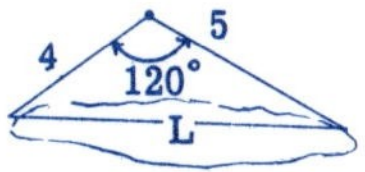

By the law of cosines,

$$L^2 = 4^2 + 5^2 - 2 \cdot 4 \cdot 5 \cos 120^\circ = 16 + 25 + 20 = 61.$$

$\therefore L = 7.81$ miles.

18.27 OBLIQUE TRIANGLE SOLUTION

A triangular plot of land has a frontage of 90 feet on one street and 120 feet on another. Find the angle the two streets make with one another if the third side of the plot is 60 feet.

18.27

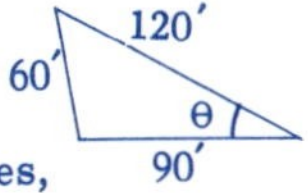

By the Law of Cosines,

$$\cos\theta = \frac{120^2 + 90^2 - 60^2}{2\cdot 120\cdot 90} = \frac{12^2 + 9^2 - 6^2}{2\cdot 12\cdot 9} = \frac{189}{2\cdot 12\cdot 9} = \frac{7}{8}$$

$\theta = 29°$

19.1 FORMULA: PARALLELOGRAM AREA

What is the area of a parallelogram with sides a and b and included angle θ?

19.1

$$K = ab \sin \theta .$$

19.2 FORMULA: QUADRILATERAL AREA

What is the area of a quadrilateral with diagonals a and b and the angle between them θ?

19.2

$$K = \frac{1}{2} ab \sin \theta$$

19.3 FORMULA: POLYGON AREA

What is the area of a regular polygon with n sides each of length L?

19.3

$$K = \frac{1}{4} n L^2 \cot \frac{180^\circ}{n}$$

19.4 FORMULA: POLYGON AREA

Find the area of a 6 sided polygon, each side being of length 10.

19.4

$$K = \frac{1}{4} \cdot 6 \cdot 100 \cdot \cot \frac{180^\circ}{6} = 150 \cdot \cot 30^\circ = 150 \cdot \sqrt{3}\,.$$

19.5 FORMULA: INSCRIBED CIRCLE

What is the radius of a circle inscribed in a regular polygon of n sides, each of length L?

19.5

$$r = \frac{L}{2} \cot \frac{180^\circ}{n}$$

19.6 FORMULA: INSCRIBED CIRCLE

Find the radius of a circle inscribed in a regular 6-sided polygon, each side of length 10.

19.6

$$r = \frac{10}{2} \cot \frac{180^\circ}{6} = 5 \cot 30^\circ = 5 \cdot \sqrt{3}$$

19.7 FORMULA: CIRCUMSCRIBED CIRCLE

What is the radius of the circle circumscribed about a regular polygon of n sides, each of length L?

19.7

$$R = \frac{L}{2} \csc \frac{180^\circ}{n}$$

19.8 FORMULA: CIRCUMSCRIBED CIRCLE

Find the radius of the circle circumscribed about a regular 6-sided polygon, each side of length 10.

19.8

$$R = \frac{10}{2} \csc \frac{180^\circ}{6} = 5 \csc 30^\circ = 5 \cdot 2 = 10$$

19.9 FORMULA: INSCRIBED CIRCLE

What is the radius of the circle inscribed in any triangle with sides a, b and d?

19.9

$$r = \frac{\sqrt{s(s-a)(s-b)(s-c)}}{s} \quad \text{where} \quad s = \frac{1}{2}(a+b+c)$$

19.10 FORMULA: INSCRIBED CIRCLE

Find the radius of the circle inscribed in a triangle with sides 5, 12 and 13.

19.10

$$s = \frac{1}{2}(5 + 12 + 13) = 15$$

$$r = \frac{\sqrt{15 \cdot 10 \cdot 3 \cdot 2}}{15} = \frac{\sqrt{900}}{15} = \frac{30}{15} = 2$$

19.11 FORMULA: CIRCUMSCRIBED CIRCLE

What is the radius of the circle circumscribed about any triangle with sides a, b and c?

19.11

$$R = \frac{abc}{4\sqrt{s(s-a)(s-b)(s-c)}}$$

where $s = \frac{1}{2}(a+b+c)$

19.12 FORMULA: CIRCUMSCRIBED CIRCLE

Find the radius of the circle circumscribed about the triangle with sides 5, 12 and 13.

19.12

$$s = \frac{1}{2}(5+12+13) = 15$$

$$R = \frac{5\cdot12\cdot13}{4\sqrt{15\cdot10\cdot3\cdot2}} = \frac{5\cdot12\cdot13}{4\sqrt{900}} = \frac{5\cdot12\cdot13}{4.30} = \frac{13}{2}$$

19.13 FORMULA: POLYGON PERIMETER

What is the perimeter of a polygon of n sides inscribed in a circle of radius r?

19.13

$$P = 2nr\sin\frac{\pi}{n}$$

19.14 FORMULA: POLYGON PERIMETER

What is the perimeter of a five-sided regular polygon inscribed in a circle of radius 10?

19.14

$$P = 2 \cdot 5 \cdot 10 \sin\frac{\pi}{5} = 100 \sin 36^\circ = 58.8$$

19.15 FORMULA - POLYGON AREA

What is the area of a polygon of n sides inscribed in a circle of radius r?

19.15

$$K = \frac{1}{2} n r^2 \sin \frac{2\pi}{n}$$

19.16 FORMULA: POLYGON AREA

What is the area of a regular five-sided polygon inscribed in a circle of radius 10?

19.16

$$K = \frac{1}{2} \cdot 5 \cdot 10^2 \cdot \sin\frac{2\pi}{n} = 250 \sin 72^\circ = 237.8$$

19.17 FORMULA: CIRCUMSCRIBED POLYGON

What is the perimeter of a polygon of n sides circumscribed about a circle of radius r?

19.17

$$P = 2\,n\,r\tan\frac{\pi}{n}\ .$$

19.18 FORMULA: CIRCUMSCRIBED POLYGON

What is the perimeter of a regular 5-sided polygon circumscribed about a circle of radius 10?

19.18

$$P = 2.5\,(10 \tan {}^{\pi}/5)$$
$$= 100 \tan 36^\circ$$
$$= 72.7$$

19.19 FORMULA: CIRCUMSCRIBED POLYGON

What is the area of a polygon of n sides circumscribed about a circle of radius r?

19.19

$$K = n r^2 \tan \frac{\pi}{n}$$

19.20 FORMULA: CIRCUMSCRIBED POLYGON

What is the area of a regular 5-sided polygon circumscribed about a circle of radius 10?

19.20

$$K = 5 \cdot 10^2 \cdot \tan\frac{\pi}{n} = 500 \cdot \tan 36^\circ = 363.5$$

19.21 FORMULA: CHORD OF CIRCLE

What is the length of a chord of a circle of radius r subtending an angle θ?

19.21

$$L = 2\,r\,\sin\frac{1}{2}\theta.$$

19.22 FORMULA: CHORD LENGTH

What is the length of a chord of a circle of radius 10, which subtends an angle ($\angle$) $\theta = 50°$?

19.22

$$L = 2 \cdot 10 \sin\frac{50^\circ}{2} = 20 \ \sin 25^\circ = 8.45$$

19.23 FORMULA: SECTOR OF CIRCLE

What is the area of a sector of a circle included between two radii having an angle θ between them?

19.23

$$K = \frac{sr}{2} = \frac{1}{2}r^2\theta$$

where $s = r\theta$ and θ is in radians.

19.24 FORMULA: SECTOR OF CIRCLE

What is the area of a sector of a circle of radius 10 included between two radii having an included angle of 50°?

19.24

$$K = \frac{1}{2}(10^2) \cdot 50 \cdot \frac{\pi}{180} = \frac{125\pi}{9}$$

19.25 FORMULA: SEGMENT OF CIRCLE

What is the area of a segment of a circle (the area included between an arc of length s and the corresponding chord) where θ is the angle included between the two radii.

19.25

$K = \frac{1}{2} r^2 (\theta - \sin\theta)$ where θ is in radians.

19.26 FORMULA: SEGMENT OF CIRCLE

What is the area of a segment of a circle of radius 10, if the two radii of the circle have an included angle of 150°?

19.26

$$K = \frac{1}{2} r^2 (\theta - \sin\theta) = \frac{1}{2} \cdot 10^2 \left(120 \cdot \frac{\pi}{180} - \sin 150^\circ\right)$$

$$= 50\left(\frac{2\pi}{3} - \frac{1}{2}\right) = 50\left(\frac{4\pi - 3}{6}\right) = \frac{25}{3}(4\pi - 3)$$

20.1 GRAPHS OF TRIG. FUNCTIONS: PERIODIC

What is meant by a periodic function?

20.1

Let $f(x)$ be a given function. Then, if there is a real number k, different from 0, such that $f(x+k) = f(x)$, then $f(x)$ is called periodic with period k.

20.2 GRAPHS OF TRIG. FUNCTIONS: PERIOD

What is the period of the function sin x?

20.2

As we may write $\sin(x + 2\pi) = \sin x$,

by definition the period of $\sin x$ is $k = 2\pi$.

20.3 GRAPHS OF TRIG. FUNCTIONS: PERIOD

What is the period of the function cos x?

20.3

As we may write $\cos(x + 2\pi) = \cos x$; by definition, the period of $\cos x$ is $k = 2\pi$.

20.4 GRAPHS OF TRIG. FUNCTIONS: PERIOD

What is the period of the function tan x or cot x?

20.4

As $\tan(x+\pi) = \tan x$

$\cot(x+\pi) = \cot x$,

the period of both $\tan x$ and $\cot x$ is $k = \pi$.

What are the essential elements of a graph of a trigonometric function?

20.5

The essential elements are 1) the period, as defined in card 20.1, and 2) the amplitude, which measures the maximum ordinate, or distance above or below the horizontal axis.

20.6 GRAPHS OF TRIG. FUNCTIONS - PERIOD AND AMPLITUDE

How are the period and amplitude of the functions $y = a \sin b\,x$ or $y = a \cos bx$ obtained?

20.6

The period is $\frac{2\pi}{b}$; the amplitude is a.

20.7 GRAPH OF TRIG. FUNCTIONS: PERIOD AND AMPLITUDE

How are the period and amplitude of the functions y = a tan bx and y = a cot bx obtained?

20.7

The period is $\frac{\pi}{b}$; the amplitude is infinite.

20.8 CIRCULAR FUNCTIONS

Why are the trigonometric functions called circular functions?

20.8

If a circle of radius one is constructed, with center at the origin, all functions of a central angle of the circle can be shown as line segments, as follows:

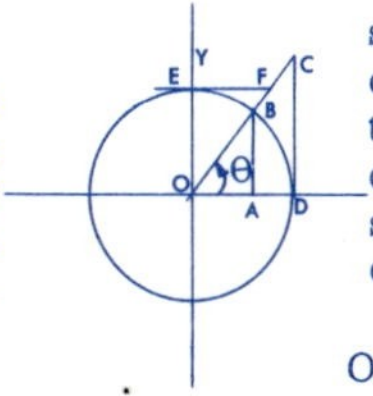

$\sin\theta = AB$
$\cos\theta = OA$
$\tan\theta = DC$
$\cot\theta = EF$
$\sec\theta = OC$
$\csc\theta = OF$

$OD = OE = 1$

20.9 GRAPH OF TRIG. FUNCTIONS - SIN

Show a graph of the function $y = \sin x$ for two periods (from $x = -2\pi$ to $x = +2\pi$).

20.9

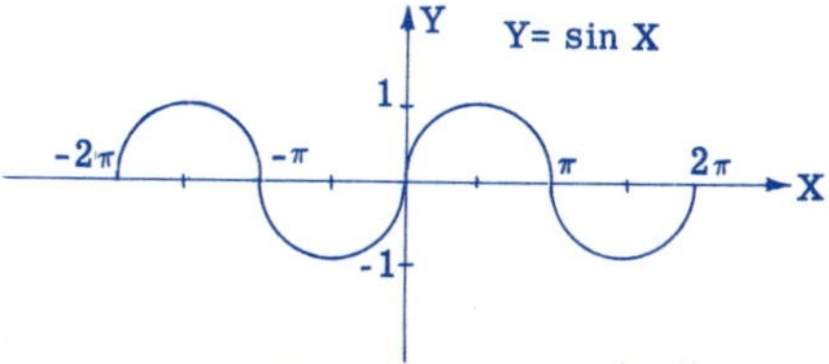

20.10 GRAPH OF TRIG. FUNCTIONS - COS

Show a graph of the function $y = \cos x$ for two periods (from $x = -2\pi$ to $x = +2\pi$).

20.10

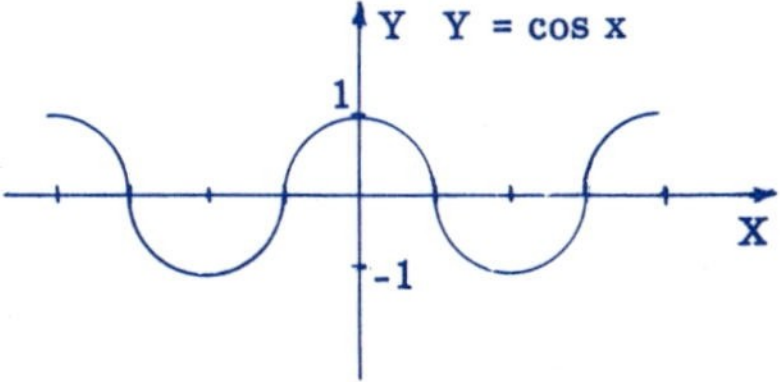

20.11 GRAPH OF TRIG. FUNCTIONS - TAN

Show a graph of the function $y = \tan x$ for two periods (from $x = -\pi$ to $x = +\pi$).

20.11

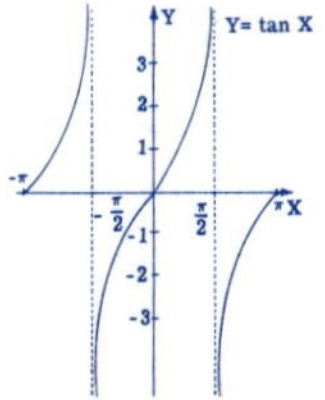

Find the period and amplitude of the function $y = 3 \sin \pi x$ and plot for $-1 \leq x \leq 3$.

20.12

$$\text{Period} = \frac{2\pi}{\pi} = 2$$

$$\text{Amplitude} = 3$$

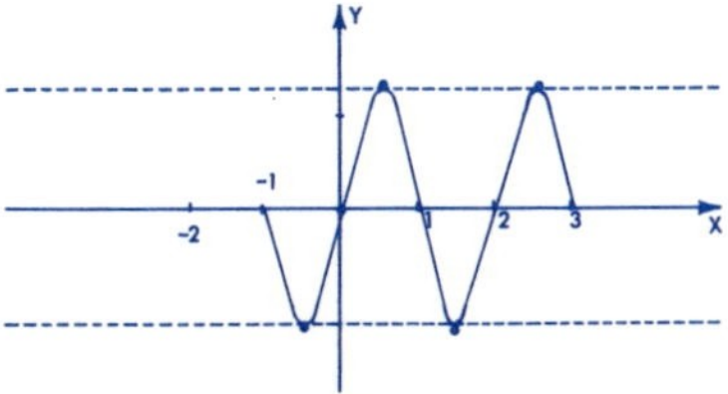

20.13 GRAPHS OF TRIG. FUNCTIONS - PERIOD AND AMPLITUDE

Find the periods and amplitudes of the following functions:

1) $y = \frac{3}{2}\cos\frac{1}{3}x$

2) $y = 2\sin 2x$

3) $y = 2\tan \pi x$

20.13

	Period	Amplitude
1)	$\frac{2\pi}{1/3} = 6\pi$	$\frac{3}{2}$
2)	$\frac{2\pi}{2} = \pi$	2
3)	$\frac{\pi}{\pi} = 1$	∞

Graph the function $y = \sin^{-1} x$ for its principal values.

20.14

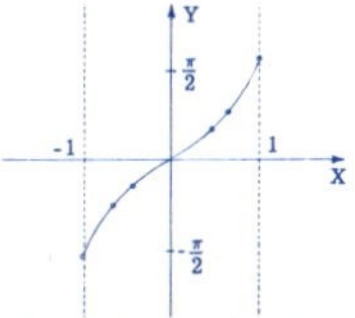

As $y = \sin^{-1} x$ is equivalent to $x = \sin y$, this becomes a sine curve with the amplitude the abscissa or x value.

20.15 GRAPHS OF INVERSE TRIG. FUNCTIONS

Graph the function $y = \cos^{-1} x$ for its principal values.

20.15

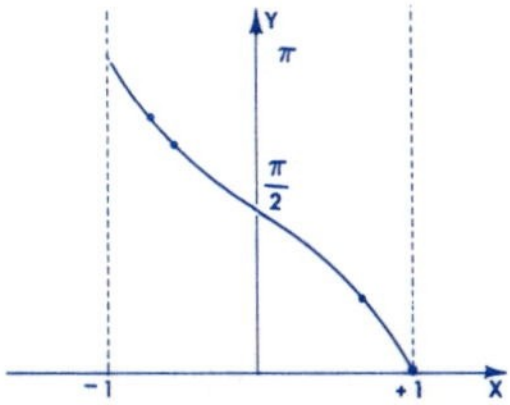

21.1 CONSTRUCTION: PERPENDICULAR BISECTOR

Construct the perpendicular bisector of a given line segment.

21.1

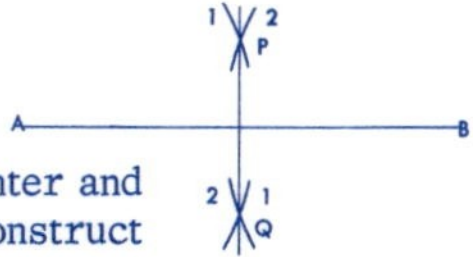

Given AB, (1) with A as center and with any convenient radius, construct arcs, labeled 1 (2) do likewise with B as center, labeled 2. These arcs intersect at P and Q. Then PQ is the desired line.

21.2 CONSTRUCTION: PERPENDICULAR LINE

Construct a perpendicular to a given line through a given point on the line.

21.2

AB is the given line, P the given point. 1) With P as center and any radius, cut AB at C and D. 2) With C as center and radius >CP, construct an arc. Do likewise with D as center, using same radius. 3) These arcs intersect at Q. Then PQ is the desired line.

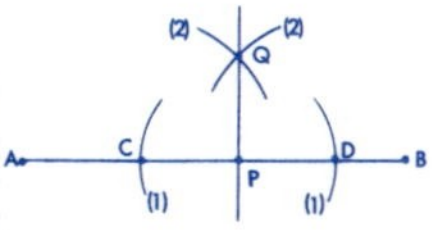

21.3 CONSTRUCTION: ANGLE BISECTOR

Construct the bisector to a given angle.

21.3

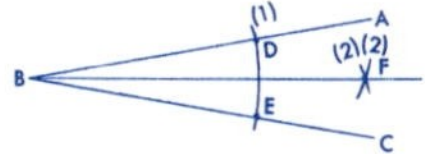

∠ABC is given.

1) With B as center and with any radius, cut AB at D and BC at E.
2) With D and E as centers and with equal radii, draw arcs intersecting at F. Then BF is the desired line.

21.4 CONSTRUCTION: EQUAL ANGLE

Construct an angle equal to a given angle at a given point on a given straight line.

21.4

Given ∠ABC: to construct an ∠equal to it at point D on line DE. 1) With B as center, draw arc FG. 2) With D as center, and same radius, draw arc MN. 3) With N as center, and FG as radius, cut arc MN at P. Then ∠ PDN is required angle.

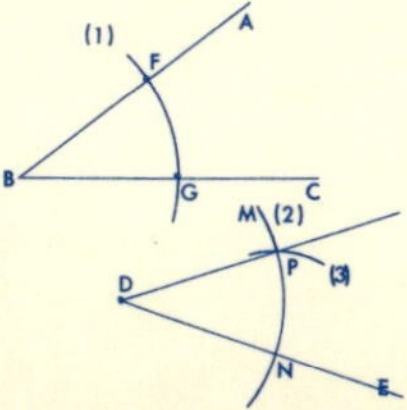

21.5 CONSTRUCTION: PARALLEL LINE

Construct a line parallel to a given line through a given point not on the line.

21.5

Given line AB and point C.

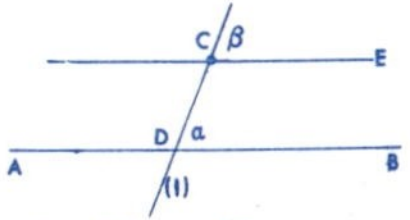

1) Draw a line through C to any point D on AB.
2) Construct CE so that $\angle a = \angle \beta$. Then CE is the desired line.

21.6 CONSTRUCTION: DIVISION OF LINE SEGMENT

Divide a given line segment into three equal parts by construction. (This can be done for any number of equal parts.)

21.6

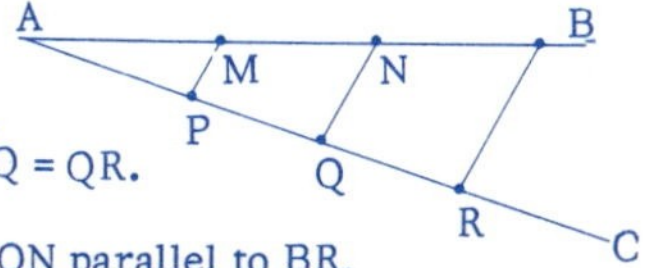

Given line AB.

1) Draw any line AC.
2) Construct AP = PQ = QR.
3) Construct BR.
4) Construct PM and QN parallel to BR.
   Then AM = MN = NB.

21.7 CONSTRUCTION: CIRCUMSCRIBED CIRCLE

Construct a circle which circumscribes a given triangle.

21.7

Given triangle ABC

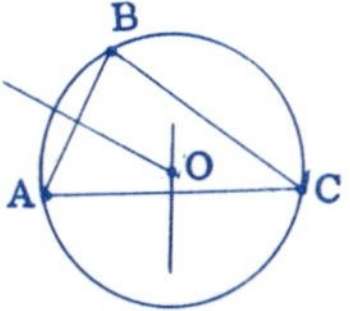

1) Draw the ⊥ bisector of sides AB and AC. These intersect at O. 2) With OA as radius and center at O, construct the required circle.

21.8 CONSTRUCTION: INSCRIBED CIRCLE

Construct a circle which is inscribed in a given triangle.

21.8

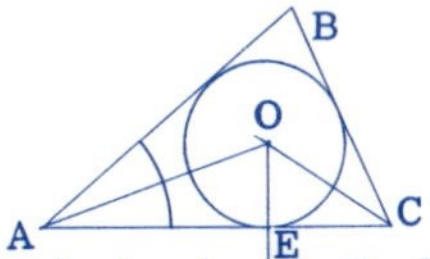

A B C is the given triangle.

1) Construct the bisectors of angle A and angle C. They meet at a point O. 2) From O, construct O E ⊥ AC. 3) With O as center and radius O E, construct the inscribed circle.

21.9 CONSTRUCTION: TANGENT TO CIRCLE

Construct a tangent to a given circle from a point exterior to the circle.

21.9

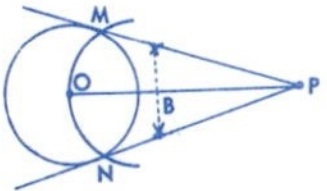

Given circle O and exterior point P.
1) Construct OP. 2) Bisect OP at B. 3) With B as center and radius OB, draw arcs cutting circle O at M and N. 4) Then PM and PN are the tangents.

21.10 CONSTRUCTION: TRIANGLE

Construct a triangle given its three sides.

21.10 a, b, c, are given sides

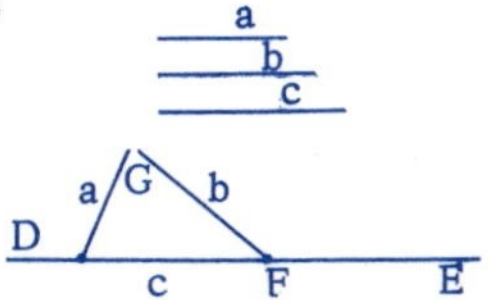

1) With D as center, construct DF = c. 2) With D as center, construct an arc = a. With F as center, construct an arc = b. These arcs intersect at G. Then DFG is desired triangle.

21.11 CONSTRUCTION: TRIANGLE

Construct a triangle given two sides and the included angle.

21.11

Angle A and sides b and c are given.

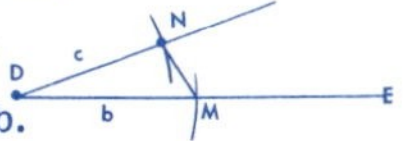

1) With D as center, construct DM = b.
2) Construct ∠D = ∠A
3) With D as center, construct DN = c. Then △DNM is the desired triangle.

21.12 CONSTRUCTION: TRIANGLE

Given two angles whose sum is less than a straight angle, and the included side; construct a triangle.

21.12

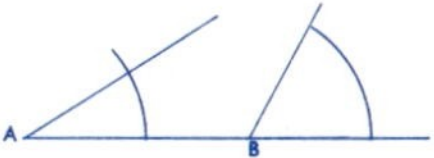

Given angles A and B and side c.

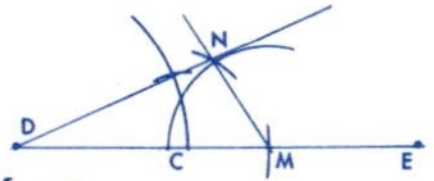

1) With D as center, construct DM = c.
2) Construct ∠D = ∠A; at M construct an angle equal to B. Then DMN is the required triangle.

21.13 CONSTRUCTION: TRIANGLE

Given two angles whose sum is less than a straight angle, and a side opposite one of them, construct a triangle.

21.13

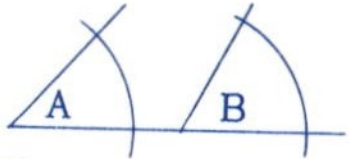

Given angles A and B and side a

1) With D as center, construct DM = a
2) At D construct an angle equal to ∠B.
3) At M construct an angle equal to ∠A +∠B.
4) Then DMN is the required triangle.

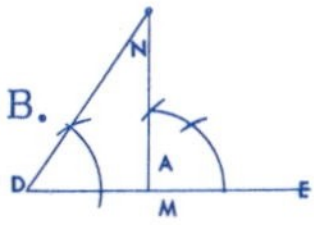

21.14 CONSTRUCTION: MEAN PROPORTIONAL

Construct a mean proportional to two given line segments.

21.14

a and b are the given line segments

1) On AE, take AB = a and BC = b. 2) Bisect AC at F. 3) With F as center and FA as radius, construct a semicircle. 4) At B, construct BD ⊥ AC. 5) Then BD is the required mean proportional or by similar △

$$\frac{AB}{BD} = \frac{BD}{BC}$$

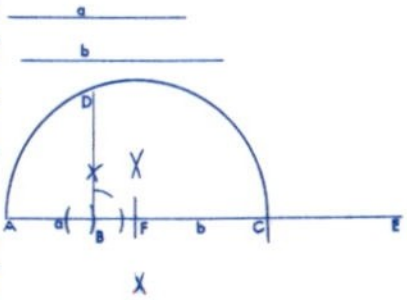

21.15 CONSTRUCTION: SQUARE ROOT

Find by construction, the square root of a number.

21.15

By the previous construction, let $a = 1$ and b be the representation of the number. Let $BD = x$.

Then $\frac{1}{x} = \frac{x}{b}$; $x^2 = b$; $x = \sqrt{b}$.

21.16 CONSTRUCTION: SQUARE OF NUMBER

Find by construction, the square of a number.

21.16

Given b as the representation of the number, and a unit length 1. 1) Construct DM = 1. 2) Construct MN ⊥ DE and equal to b. 3) At N construct NP ⊥ DN. 4) Then MN = b, as by similar triangles, $\frac{1}{b} = \frac{b}{MP}$ or $MP = b^2$.

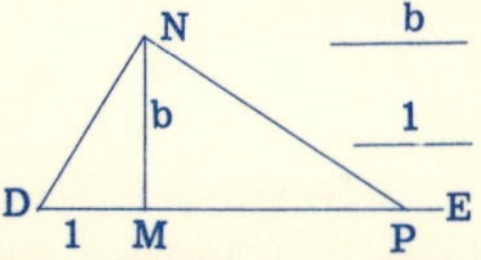